Anonymous Men
of the Old Testament

God's Anonymous Collection
Volume II

Clarisse Barros

INTERNATIONAL

PETRADI
ΠΕΤΡΑΔΙ

PUBLISHING HOUSE

Anonymous Men of the Old Testament

Original Portuguese Copyright © 2024 by Clarisse Barros
English translation © 2025 by Petradi International Publishing House, LLC
All rights reserved.

Petradi International Publishing House books may be purchased in bulk at special discounts for sales promotion, corporate gifts, ministry, fund-raising, or educational purposes. Special editions can also be created to specifications. For details, contact Special Sales Dept., Petradi International Publishing House at info@petradipublishing.com.

Visit our website at www.petradipublishing.com.

Library of Congress Control Number:

ISBN: 978-1-968609-02-3
eBook ISBN: 978-1-968609-03-0

Cover Design by Dan Pitts, Dan Pitts LLC

Printed in the United States

10 9 8 7 6 5 4 3 2 1

Dedication

To my husband,
my son,
my son–in–law,
my grandchildren and my father.

To the Lord Jesus Christ – God made Man.

Acknowledgements

To the Lord, for showing me that writing is a pulpit, and that I have a ministry to fulfill.

To Carlos Cunha, for coordinating and editing this volume. To CLC–Coimbra, for publishing this work.

To Bruno Pires, paginator.

To Susana Pinho, my Portuguese proofreader, for her tireless and excellent work.

To Beatriz Leite, for her valuable collaboration in the field of mathematics.

To Ana Ester Tavares, my daughter, for everything she has taught me in the field of music.

To Bruno Pires, for the magnificent illustrations and the cover.

To Dr. Palmeiro Barros, my husband, who knows that I write for hours on end, usually during the day, but often at dawn, when the silence is greater, and who always supports me.

To those who pray tirelessly for my writing ministry (your prayers have made all the difference!).

To my readers, for walking with me through the pages of Scripture.

My greatest gratitude!

Contents

Introduction

This is the second volume in the series God's Anonymous people. The first one tells us about women from the Old Testament whose names are not revealed to us, but who, despite their anonymity, were instruments of great achievements, such as the Egyptian princess who saved the baby Moses from the waters of the Nile, or the mother of Samson, who gave birth to a special boy, a son like no other, endowed with extraordinary physical strength, chosen by God to be the judge of his people for twenty years. Like her, many other "nameless" women have their faith, their courage and their example of life to follow, like Noah's wife, or to avoid, like Lot's wife, recorded in the Bible, even though we don't know their names.

Over the years, I have always appreciated the courage of so many people to go unnoticed in the world, to do simple but important things over and over again, without expecting any recognition, to carry out the mundane tasks of everyday life, without which life for all of us would be very difficult, if not impossible!

I thank God for all the people who help make our society work, from the baker who gets up at dawn to knead our bread, to the truck driver who drives his truck for hours so that we have access to the products we need. To the garbage collectors, street sweepers, laborers, miners, teachers and educational assistants, doctors, and all the staff who work hard to ensure that we have decent, quality health care when we need it (especially during this pandemic); the miners, the builders, the train drivers, the pilots, the administrators, the researchers, the scientists, the engineers, the architects, the poets, the painters, the musicians, the dreamers, and all the many people who never make the headlines but without whom the world would be a much worse place!

The Bible mentions the deeds and words of people whose names are not recorded along with their accomplishments. But God, the Almighty, who chooses and empowers individuals for various projects and tasks, knows who they are!

Your name dear, reader, is in God's records! In His books are recorded the efforts you've made, the use you've made of your gifts and talents, every drop of sweat, every tear that no one noticed (except God!), every failure and every victory, what you learned from the difficulties you encountered along the way, how you applied your knowledge, how you developed those muscles of faith and that firm warrior stance! God values your story. God knows your heart. God knows what no one else knows about you because He has always kept His eyes on you!

One day your name will be known. How do you want to be known? What words or titles would you like to have associated with your name? dear Readers, the choice is yours!

My prayer is that this book you are holding in your hands will change the way you see yourself; that it will change the way you see God's purposes and the mission He has carefully chosen for you; that it will change the way you look at your job, your financial situation, your position in the company where you work, the opportunities that come your way, the way you see hierarchy; that it will change the way you see and value your family, your mission as a husband, father, son, uncle, grandfather, friend, neighbor, colleague, or any other role. I pray that this book will show you the value and eternal reach of your simplest gestures, words, and attitudes. Finally, I pray that you will see the love that God has for you and that you will learn more about the extent of that inexhaustible love!

God bless you mightily!

Clarisse Barros

Part I

Men Builders

Chapter 1
Noah's Carpenters

- Historical data -

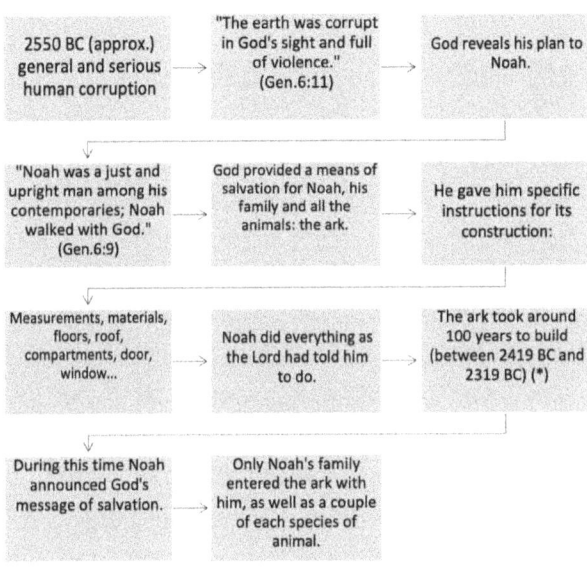

2550 BC (approx.) general and serious human corruption	"The earth was corrupt in God's sight and full of violence." (Gen.6:11)	God reveals his plan to Noah.
"Noah was a just and upright man among his contemporaries; Noah walked with God." (Gen.6:9)	God provided a means of salvation for Noah, his family and all the animals: the ark.	He gave him specific instructions for its construction:
Measurements, materials, floors, roof, compartments, door, window...	Noah did everything as the Lord had told him to do.	The ark took around 100 years to build (between 2419 BC and 2319 BC) (*)
During this time Noah announced God's message of salvation.	Only Noah's family entered the ark with him, as well as a couple of each species of animal.	

(*) According to the Bible in Chronological Order by Editora Vida.

God behind the scenes

Everyone's life has a story. Regardless of your name and the time in which each of us lives, God is interested in being a part of that story.

He creates each human being with the holy intention of "intertwining" and "intervening" in their lives. He tries to make himself heard and understood in many ways. Just look at some of them with me:

- If we look closely, we can easily discover God's fingerprints in the wonders of creation – in the grandeur of a mountain, in the perfection of created beings, each with their own unique characteristics, their knowledge, their purpose, their way of living and communicating. We will also discover it in an example of faith, in the character of a person shaped by his Creator...
- We can also see God's signature in visible and invisible, microscopic things whose existence has been discovered by the persistent research of various sciences;
- If we listen carefully, we will hear the whisper of His voice both in the passing wind and in the silence of the forest; in the lapping of the waves on a calm day or in the rumble of thunder on a stormy day; in our conscience, in our heart; and finally, clearly and unequivocally, in His written Word!
- In addition to all this, there are circumstances in life that point to God: a deliverance, the solution to a complex problem, a unique opportunity that presents itself to us, a hand that holds ours in a difficult moment, a phone call from a friend who hasn't called in a long time, help that comes to us unexpectedly.

God moves for us. He thinks of us. He loves us always, forever. He is writing our story. We are not alone on the stage of life. The writer of the script is acting with us!

We will read a little about the story of Noah and how God intervened in his life in the following Bible passage (in the New Living Translation).

"This is the account of Noah and his family. Noah was a righteous man, the only blameless person living on earth at the time, and he walked in close fellowship with God. Noah was the father of three sons: Shem, Ham, and Japheth. Now God saw that the earth had become corrupt and was filled with violence. God observed all this corruption

in the world, for everyone on earth was corrupt. So God said to Noah, "I have decided to destroy all living creatures, for they have filled the earth with violence. Yes, I will wipe them all out along with the earth!

Build a large boat from cypress wood and waterproof it with tar, inside and out. Then construct decks and stalls throughout its interior. Make the boat 450 feet long, 75 feet wide, and 45 feet high. Leave an 18-inch opening below the roof all the way around the boat. Put the door on the side, and build three decks inside the boat—lower, middle, and upper.

"Look! I am about to cover the earth with a flood that will destroy every living thing that breathes. Everything on earth will die. But I will confirm my covenant with you. So enter the boat—you and your wife and your sons and their wives. Bring a pair of every kind of animal—a male and a female—into the boat with you to keep them alive during the flood. Pairs of every kind of bird, and every kind of animal, and every kind of small animal that scurries along the ground, will come to you to be kept alive. And be sure to take on board enough food for your family and for all the animals."

So Noah did everything exactly as God had commanded him. (Genesis 6:9–22 NLT)

Noah's carpenters - did they exist?

When Noah received the extraordinary mission from God to build the great ark, he had to calculate the materials that would be needed for the project. It wasn't his project. It was God's project! But he, as a man, would have the high privilege of carrying it out.

God has plans and projects that no one but you can bring to fruition, in God's way, in God's time! Think about it!

Noah's working skills, knowledge, and experience were not enough. Even if he added all the talents of his three sons, all his strength, skills and abilities, all the good ideas of his wife and three daughters–in–law, he would still be far from having the knowledge, strength and expertise needed to build the ark! There were so many factors to consider! Noah would need help. He needed to sit down and do the math. Shall we go into numbers? Here are some suggestions:

- How many trees would have to be cut down in the forest?
- How many boards, planks, beams, and slats of wood were needed?
- How many nails?
- How many hammers?
- How many saws, knives, and blades of different cuts?
- How much bitumen to insulate the big boat?
- How much power?
- How many men?
- How many carpenters?
- How many blacksmiths would be needed to keep the tools sharp and in good working order?
- How long would it take just to prepare the materials?
- How much time did he have each day to do the job?
- What was the deadline for completion?
- How many partitions would have to be made on the three floors of the ark ?
- How much water would they need to store? How much food? For how long?
- How many animals were coming? He didn't know all the species... There was no encyclopedia to enlighten him... Only God.

When we are called to a mission, whatever it may be, it is important to sit down and do the math, as the Lord Jesus Christ taught:

"But don't begin until you count the cost. For who would begin construction of a building without first calculating the cost to see if there is enough money to finish it? – Luke 14:28 "(...) to see if there is enough money to finish it (...)" means to see if you have the resources to start, continue, and finish the project in a way that honors God. To see if you have the faith to go forward, the foundation confirmed by God's Holy Spirit and His Word, the financial and human resources, if necessary.

When we look at a finished project, we don't have a clear idea of everything that happened during its construction. We can't accurately assess the costs, the material and human resources, the effort, the discipline, the perseverance, the injuries and accidents on the job, the sacrifices, the doubts that arose, the problems that arose and the solutions that were found to resolve them, what was left behind to ensure that the project was completed and delivered on time... So many details that completely escape us!

Did Noah try to find workers and helpers? Maybe some carpenters or lumberjacks? Maybe some strong and courageous friends? The Bible doesn't tell us what decisions were made regarding these aspects. There is, however, something that is clearly conveyed to us:

God "(...) did not spare the ancient world, but preserved Noah, a preacher of righteousness, with seven others, when He brought a flood upon the world of the ungodly..." (2 Peter 2:5)

God preserved Noah and seven other people–his family. Noah was a preacher: He spoke of God's justice that would come upon the world if men and women didn't repent of their sins, if they didn't stop doing wrong, if they didn't turn to their Creator, If they didn't flee into the only means of salvation that would be made available – the great boat that became known in human history as Noah's Ark!

God preserved lives and provided the means of escape, the safe haven for all who would receive His message of righteousness and judgment. God made Noah a fearless preacher! He gave him a hard but realistic message to preach and time for everyone to hear it.

To whom did Noah announce God's plan?

Come on, reader, let's look at the curious people who came to see what Noah was doing:

As usual when we start a work for God, critics came (and still come today) from all sides! They came with a huge list of provocative questions, hurling them at Noah and his children as if they were pointed stones or *darts fired by the evil one!*

Have you ever had such an experience? If so, you fully understand the pain of opposition. I invite you to keep your eyes fixed on God's plan for your life. Do as Noah did! Keep going, no matter what anyone says!

Noah's critics, a crowd of anonymous accusers, came with heavy burdens of discouragement and unloaded them on Noah, his three sons, and their wives. It may have been something similar to what you're about to read:

- You're crazy, Noah!
- Stop it!
- Stop this nonsense at once! Where have you seen a boat built so far from the sea?
- The work is not well done!
- The wood is weak!
- Even if you cut down all the forests in the world, you'll never have enough sturdy wood for this endeavor!
- When the flood comes, your boat won't be able to handle it...
- How can you be sure that this flood will come?
- How do you know that it was God, the Creator, who spoke to you? Wasn't it a dream? Haven't you imagined all this? Isn't it all in your head?
- If you ever finish your construction... No, you'll never finish it.
- It's too big! You'll never get it to the sea!
- No! It's too small! It's not enough for anything! You say the animals will come from all over the world... in pairs!
- I'd even help you from time to time if I could, because you're a good man... But I'm not crazy! I'm not going to get into something like that just so everyone can laugh at me!
- Give you a hand? I have more to do! My hands aren't big enough for my job, let alone yours!
- I'm really busy. I don't have time.
- You are in big trouble!
- You got yourself into trouble, now get out of it!
- I might stop by to see what you're up to, but don't think I'll be working!

How difficult! When we are criticized by people who ridicule us and try to discredit our character and our work, there is only one way to win: Forgive, forgive, forgive again, and continue to listen to the voice of God in our hearts, the word that the Holy Spirit brings to our minds, the confirmation that we are doing the right thing.

Now let's imagine that Noah is trying to recruit carpenters. Turning to one of the curious people he knew well, he might say:

Noah: I could use your expertise...

Carpenter: Forget it! I'll go, but no strings attached. I have my life.

Noah: When God fulfills His promise, you will no longer have a life, a home, a family, a job...

Carpenter: Is that a threat?

Noah: No! It's a warning from God!

Carpenter: Yeah? Well, tell him I'm not interested!

Have you ever met such people, dear reader? People who are curious about your life and about God, but don't want to make any commitments, don't want to be identified as believers, let alone be associated with "works of God", churches, organizations, groups... One day each of these people will have a glimpse of the depth of God's wisdom and His immeasurable love! May it not be too late, as it was for the carpenters who could have entered the ark with Noah, but chose not to.

Noah probably encountered several types of possible associates. Let's see:

A shy carpenter (like Nicodemus, the man who sought out Jesus at night so as not to be seen in His company, as we read in John chapter 3, and ended up hearing the most famous words in the entire Bible – John 3:16):

- I will help you, I really want to know what is going to come out of this building, but I don't want it to be known that I was here! I'll come at dusk...

An occasional carpenter (like the Samaritan woman who found Jesus tired, sitting by Jacob's well; who heard him speak of living water, water that coulf quench human thirst forever that she thought about "taking advantage" of the offer of this water, according to the account in John chapter 4. In the end, she understood who was speaking to her, she found forgiveness for her sins, and her life was transformed by this encounter with Jesus, who revealed Himself to her as the Christ):

- I feel sorry for you, Noah, you are exhausted! I'll help you cut down some trees. We could use some clearing in this area. We'll all benefit from it...

An unbelieving carpenter (like the man who asked Jesus to heal his sick son, as we read in Mark chapter 9, while finding it extremely difficult to believe that the Lord could do anything for him! He asked Him to help him in his unbelief and watched in amazement as his son was healed):

- You're been dwelling on this idea of the flood and the boat for so long! It could have been worse for you... I don't believe any of this! I don't know how anyone could believe it! It's all in your mind!

A carefree carpenter (like the rich man who had such a disastrous harvest that he thought it necessary to build new and bigger barns to store everything he had! And so he made plans to rest, eat, drink, and have fun, forgetting that the end of his life was closer than he thought):

- Why don't you do what everyone else does? You could enjoy life, eat and drink to your heart's content and then we'll see! That's what we all do!

An envious carpenter (like the anonymous crowd that wouldn't let Zacchaeus, the short tax collector, see Jesus! But, as Luke tells us in chapter 19, he persisted in his determination to see Him, hiding under the leaves of a tree on the road, up which he climbed like a little boy, and was found by Jesus, the One from whom nothing is hidden, disguised, concealed, and nothing goes unnoticed! The envy of the crowd grew even more when the Lord accompanied Zacchaeus to his house...):

- You think you're special, that the Creator is talking to you! Do you think He has nothing else to do? Do you think you deserve more than others? Do you think he looks at you and doesn't look at us?! Don't count on me!

An ominous carpenter (like Thomas, referred to in John 11:16, who thought that the twelve disciples would die, just like Jesus, if they returned to the outskirts of Jerusalem, to Bethany, where Lazarus, the Lord's friend, was very ill, and eventually died.

The Lord returned to Bethany to raise Lazarus from the dead and not to give up his life because his time had not yet come):

- You should stop this! You'll work until you die for nothing! I still think you're out of your mind, and you're dragging your family down with you!

Noah heard everything and warned everyone he could reach:

- When God's boat is ready, the animals will come two by two, all kinds of animals; my family and I will get in, the door will be shut, and the rain will come, the great rain the Lord has promised! And whoever is not in the boat will not survive...

In the crowd of Critical Carpenters, there were still some who said something similar:

- What you are saying is impossible! If the water starts to rise, we'll head for the mountains!

And Noah continued to preach:

- The water will cover even the greatest mountains... There's no way to escape except by boat.

And the crowd kept saying:

These are just your ideas to enslave us and make us work!

- We can't believe it! Only when we see it!

Noah:

- And you will! Everyone will see, but you won't live to tell what will happen...

All the people of that generation heard Noah's message, directly or indirectly. The word went from mouth to mouth and spread across the face of the populated earth at that time. Any one of these men could have believed Noah's words and joined his family's team of carpenters. But none were willing to receive his message as the Word of God. How similar our times are to that time!

God's Message to Humanity

The central message of Scripture – the salvation of the human soul through faith in the perfect, sufficient, and irreplaceable work of redemption of Christ – has reached the planet! The pandemic has been helpful in this mission: Millions of Christians have shared their faith and proclaimed the name of Jesus through the computerized means available to us in this digital age. God's love is reaching all peoples, languages and nations! It's time to tell our generation that great things are happening and greater things are to come! Get ready!

Yet we continue to meet people who refuse to accept salvation through Christ, who find excuses for themselves mentioning the tribes and peoples of the most remote regions of the world without access to the Bible or the Internet. "What will become of them?" – they ask. As if God were unfair to these people! As if the heavens didn't proclaim God's work every day! As if nature didn't constantly point to its Creator! As if anyone could be excluded from the power of the work of the cross! As if God were a respecter of people! As if the Lord forgot anyone! See what the Bible says about this:

> "The heavens are telling of the glory of God;
> And their expanse is declaring the work of His hands.
> Day to day pours forth speech,
> And night to night reveals knowledge.
> There is no speech, nor are there words;
> Their voice is not heard." Psalm 19:1–3

"For there is no partiality with God." Romans 2:11

"By faith we understand that the worlds were prepared by the word of God, so that what is seen was not made out of things which are visible." Hebrews 11:3

"As it is written: There is none righteous, not even one." Romans 3:10

The Lord Himself made it clear that He was looking for those who understand, that they cannot save themselves and who need the Savior:

"I have not come to call the righteous but sinners to repentance." Luke 5:32

And for those who are the Lord's but are aware of their weaknesses, failures, or stubbornness, the way back to the Father is through 1 John 1:9:

"If we confess our sins, He is faithful and righteous to forgive us our sins and to cleanse us from all unrighteousness." 1 John 1:9

Now or never

The very common excuse for postponing the decision to recognize oneself as a sinner and accept God's forgiveness, or to refuse that forgiveness, is clearly a terrible trap set by Satan! You see, reader, he is an enemy of God and of you, very skilled and experienced in putting stumbling blocks in the way of faith and in erecting barriers and obstacles in people's minds! We know this. The Bible teaches us and warns us of his evil intentions! Don't go there, dear reader! Don't put off a decision that will determine your eternal future! Believe in the Lord Jesus Christ and be saved! Great and wonderful events await God's children!

Embrace the faith fearlessly and boldly! Be an example to your family, your friends, and all those who only know you by sight, but who will not fail to notice your life transformation. Be a part of God's great project for mankind! Step aboard the Ark of Salvation and sail the deep seas of the Almighty's infinite and unfathomable love!

"Today, if you hear his voice, do not harden your hearts." Hebrews 4:7

Listen to God's voice calling your name right now.

- To some he will say: *"Son, your sins are forgiven. Go. From now on sin no more."* (Mark 2:5; John 8:11)
- To others he might say: *"I know where you've been wandering, like a sheep without a shepherd. But here I am now, ready to find*

you again and be your Good Shepherd, to guide you to still waters and care for you every day, in all circumstances, even as you pass through the valley of the shadow of death." (Psalm 23)

- To many, the Lord clearly says: *"Fulfill your ministry!"* (2 Timothy 4:5)
- To all of us the Lord says: *"I will never desert you, nor will I never forsake you"* (Hebrews 13:5).

A moment alone with God

I can see you, O God, when nothing happens and when
I seem to have no way out.
Even when my petition is not answered as
I had asked and imagined,
I still believe and I can see
That you'll do what's best for me, according to the plan you've laid out.
I can see you, Lord, when it's dark:
Time has passed, the world is asleep,
And I'm still waiting for the greatest miracle.
My burden is at your feet.
I won't take it anymore, it belongs to you.
And as the day slowly dawns,
I rise in the strength of knowing who you are!

I hear your voice calling my name. I don't know what you have in store for me, but I know it's better than I could have predicted! I'm ready to answer your call, to follow you through life at every moment,

And come to the end with a happy and serene soul, because it was all worth it to have listened to you calling my name!

Chapter 2
The builders of the Tower of Babel

- Historical data -

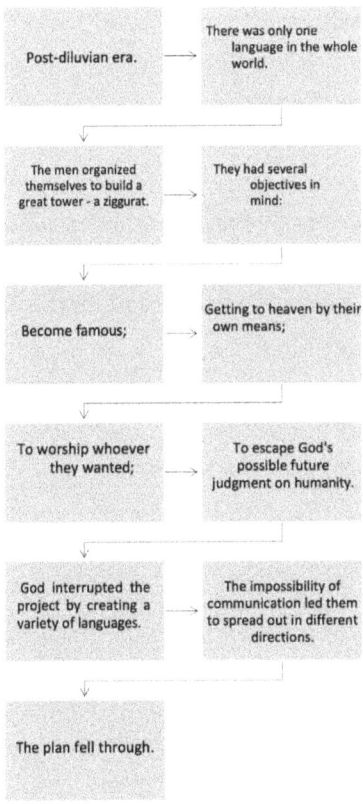

Sense of eternity

Dear reader,

Before we consider "the builders of the Tower of Babel," let's reflect for a moment on the motives that lead us to seek God's favor, to try to "get to heaven," which was the intention of those builders. Consider the text of Ecclesiastes 3:11, quoted here in two different translations:

"He has made everything appropriate in its time. He has also set eternity in their heart, yet so that man will not find out the work which God has done from the beginning even to the end." (Ecclesiastes 3:11)

"He has made everything beautiful in its time. He has also set eternity in the human heart; yet no one can fathom what God has done from beginning to end." (Ecclesiastes 3:11, NIV)

There's no denying it: all human beings are created with a sense of eternity. Since childhood, we have had the desire to live forever and to keep the people who love us – and whom we love – with us forever!

When death intervenes and "steals" a person who is part of our history, we feel an intense and immense difficulty in dealing with this loss, this absence. We basically need two certainties: (1) to know where that person has gone, and (2) to know if we will ever see them again.

On the other hand, when we deal with our own fragility and eventual departure, the questions are the same:

- Where am I going?
- Will I ever see my loved ones again?

Our *sense of eternity* tells us that we will live on after life on earth. The people we love will live on as well. The beginning of life, the birth of a new person, fragile and small, dependent on its parents, is *easy* to understand. Beginnings are beautiful and full of promise! Everything is open: all the opportunities to win and be happy, all the ventures and projects – almost everything is a possibility!

However, to say that "everything ends" with death is an idea contrary to what is inscribed in our mind, soul, heart, spirit... It doesn't make sense! Without this *sense of eternity* we wouldn't seek God the way we do. Not with the same zeal, not with the same need for answers, hope and peace.

The Word of God is full of expressions that include the concept of eternity, from the fact that our God and Lord is the Eternal God, to His promises of eternal covenant and eternal life. Look at this selection of passages from the Word, among many others that could be quoted:

"When the bow is in the cloud, then I will look upon it, to remember the everlasting covenant between God and every living creature of all flesh that is on the earth." (Genesis 9:16) – God's promise to Noah after the flood was an eternal covenant. Only an eternal God can make an everlasting covenant.

"Abraham planted a tamarisk tree at Beersheba, and there he called on the name of the LORD, the Everlasting God." (Genesis 21:33) – Abraham, called the "father of faith", related to the Lord as the Eternal God and the Almighty.

""The eternal God is a dwelling place, and underneath are the everlasting arms (...)" (Deuteronomy 33:27) – These words come from Moses, the man with whom God spoke "face to face".

The book of Psalms is filled with allusions to the Lord as the One who IS from eternity past to eternity to come. In the New Testament, both the evangelists and the apostles speak of eternal life with God, attained through faith in the atoning work of Christ. Let's see:

"For God so loved the world, that He gave His only begotten Son, that whoever believes in Him shall not perish, but have eternal life." (John 3:16)

"He who believes in the Son has eternal life; but he who does not obey the Son will not see life, but the wrath of God abides on him." (John 3:36)

"If I go and prepare a place for you, I will come again and receive you to Myself, that where I am, there you may be also." (John 14:3)

"This is eternal life, that they may know You, the only true God, and Jesus Christ whom You have sent." (John 17:3)

"For the wages of sin is death, but the free gift of God is eternal life in

Christ Jesus our Lord." *(Romans 6:23)*

"If we have hoped in Christ in this life only, we are of all men most to be pitied." *(1 Corinthians 15:19)*

"For we know that if the earthly tent which is our house is torn down, we have a building from God, a house not made with hands, eternal in the heavens." *(2 Corinthians 5:1)*

"Then we who are alive and remain will be caught up together with them in the clouds to meet the Lord in the air, and so we shall always be with the Lord." *(1 Thessalonians 4:17)*

"Now to the King eternal, immortal, invisible, the only God, be honor and glory forever and ever. Amen." *(1 Timothy 1:17)*

"in the hope of eternal life, which God, who cannot lie, promised long ages ago" *(Titus 1:2)*

"Now the God of peace, who brought up from the dead the great Shepherd of the sheep through the blood of the eternal covenant, even Jesus our Lord," *(Hebrews 13:20)*

"After you have suffered for a little while, the God of all grace, who called you to His eternal glory in Christ, will Himself perfect, confirm, strengthen and establish you." *(1 Peter 5:10)*

"And the testimony is this, that God has given us eternal life, and this life is in His Son." *(1 John 5:11)*

"These things I have written to you who believe in the name of the Son of God, so that you may know that you have eternal life." *(1 John 5:13)*

In a nutshell:

The answers to the questions at the beginning of this chapter – Where am I going? Where are you going? / Will we meet again? – have their answers in the above–mentioned texts and in others that we won't mention here, but with which the reader is probably familiar.

- Yes, there is eternal life beyond this life;
- It is attained by believing in Christ's work on the cross of Calvary, by recognizing our sinful condition and our need for forgiveness;
- That life is in Christ; where Christ is, those who believe in Him will be also;

- Yes, we will be with all our loved ones who have trusted Christ as their personal Savior.

A ziggurat: the tower of Babel Biblical story:

"At one time, the whole Earth spoke the same language. It so happened that as they moved out of the east, they came upon a plain in the land of Shinar and settled down.

They said to one another, "Come, let's make bricks and fire them well." They used brick for stone and tar for mortar.

Then they said, "Come, let's build ourselves a city and a tower that reaches Heaven. Let's make ourselves famous so we won't be scattered here and there across the Earth."

God came down to look over the city and the tower those people had built.

God took one look and said, "One people, one language; why, this is only a first step. No telling what they'll come up with next—they'll stop at nothing! Come, we'll go down and garble their speech so they won't understand each other." Then God scattered them from there all over the world. And they had to quit building the city. That's how it came to be called Babel, because there God turned their language into "babble." From there God scattered them all over the world." (Genesis 11:1–8, MSG)

After the flood, Noah's descendants multiplied on the earth.

The awesome story of the great boat built by God the Almighty to hold a few land animals of each species and Noah's own family was told from generation to generation. And the years passed. Generations followed one another. Some men gained great power and dominated the others. They built cities and divided the land among themselves. And *"At one time, the whole Earth spoke the same language."* (Genesis 11:1) Communication was simple and direct. There was no doubt about what was being said through linguistic signs. The meaning and the signifier did not vary, i.e., each concept (meaning) and its corresponding acoustic image (signifier) were the same for all inhabitants of the earth!

Note:

Scholars agree that a way to record facts and ideas existed as early as 5500 years ago. Archaeological finds from the region of Mesopotamia (present–day Iraq), namely ceramic tablets with various inscriptions, confirm the existence of writing at that time. Even before there was a systematized alphabet, sounds and their corresponding images were linked in the formation of concepts. The way of writing with images and symbols representing ideas was gradually replaced by a complex system of signs representing sounds.

But the shadow of the flood story hung over every head. People forgot the promise of an everlasting covenant that God had made with Noah and his sons. They told and retold the story, but they didn't emphasize the Lord's promises enough:

"Now behold, I Myself do establish My covenant with you, and with your descendants after you; and with every living creature that is with you, the birds, the cattle, and every beast of the earth with you; of all that comes out of the ark, even every beast of the earth. I establish My covenant with you; and all flesh shall never again be cut off by the water of the flood, neither shall there again be a flood to destroy the earth." God said, "This is the sign of the covenant which I am making between Me and you and every living creature that is with you, for all successive generations; I set My bow in the cloud, and it shall be for a sign of a covenant between Me and the earth. It shall come about, when I bring a cloud over the earth, that the bow will be seen in the cloud, and I will remember My covenant, which is between Me and you and every living creature of all flesh; and never again shall the water become a flood to destroy all flesh." (Genesis 9:9–15)

Reader, never pass over God's promises! Hold on to them with all your might, for they serve as an anchor – secure and firm – in every storm of life.

In the east, in the region of what is now Iraq, men organized and joined together to build a great city. In general, men like to build, whether it's their own homes, their careers, their families, their ministries, or anything else that is a testament to their time on earth. They feel it is important to leave a visible legacy. They don't always think that leaving a good family name for their children and a good example in life are great and indelible legacies!

"A good name is to be more desired than great wealth,
Favor is better than silver and gold." (Proverbs 22:1)

Let's go back to the building and the many decisions they made about where to build and how to work.

There was one that would determine the future of the world: in the center of the city, they would build a ziggurat!

"The ziggurats were temples that occupied the center of some ancient cities.

They were places of worship for different deities, each one dedicated to a single deity. They had a square base and were built in steps, culminating in a kind of flat platform on which the god in question was worshipped. The steps were meant to symbolize the people's invitation to the descent of their god into that temple.

Excavations and archaeological studies in ancient Mesopotamia, a vast region that includes the Tigris and Euphrates rivers and the surrounding lands, have uncovered several ziggurats, including the so–called Tower of Babel. If this tower had been completed, it would have been approximately 91 meters tall – a truly extraordinary feat at the time! It would have been the world's first skyscraper! It would have become famous, attracting many visitors from neighboring cities...

According to the Bible, the purpose of this tower was to reach the heavens, to make its builders famous, and to serve as a refuge from God's possible judgment that would scatter the people to distant lands. They wanted to stay together and feel safe, so safe that not even God could touch them!

Plans made, construction began!

No stone, no mortar and no communication

There was no stone in the area. It was a rare material. The builders were in a hurry. Inventive and creative, they found a solution to the problem of the lack of stone and mortar: they made well-fired bricks to make them more resistant, and the mortar was replaced with bitumen! Having found substitute materials, all they had to do was build!

And the tower began to rise. It rose from its square base and grew, floor by floor, step by step. The taste of victory, the pride of the name, and the audacity of this great feat could almost be felt by every enthusiastic builder!

Then the Lord came down to see the building of the city and its tower. The Lord didn't come down on the steps of the ziggurat. The Lord came down by His power to see the work that was being built. It was never to be His dwelling. It was not a temple in His honor where His name would be exalted, where prayers of supplication would be lifted up to His presence! It was a place of worship for something other than God!

The Trinity of God made their decision: Men would not be destroyed.

After the flood, there would be no more floods (fulfilling the promise made to Noah and his sons), but the building of the city and its tower would stop! No pain, no death, no injury, no sickness...

People were united by the same purpose, which was conveyed and facilitated by the existence of a single language. The Lord – simply – confused human language. He simply separated the words and the sounds that went with them. People simply stopped saying meaningful sentences to each other. Commands were no longer understandable. Confusion set in. Then chaos. Then fear. No one could explain what was happening, the builders scattered, construction stopped...

They had found an alternative to stone. It wasn't strong like stone, but it was used for construction.

They had found an alternative to mortar. It didn't adhere as well, it didn't give such a firm finish, it didn't bind the bricks as perfectly, but it was good for gluing, for holding...

But the question of language had no solution, there was no alternative! It was decided by God. And when the Lord acts, who can stand in the way? (Isaiah 43:13)

The art of communication

Several millennia have passed since the confusion of languages at Babel. However, communication between people is still not a simple matter, whether we speak the same language or not. Communication is an art that can be learned, and we all have time to develop it, to grow as communicators. Deep down, we all want to be understood and we've all thought (or said, or heard): "You don't understand me!" or "You don't get what I mean!"

Have you ever had the experience of not being understood or of not being able to understand others? When this happens, we close in on ourselves, isolating ourselves from others, from those who "don't understand" us. Then we realize that there is no silence in our thoughts. Instead, there is a constant buzz, a whirlwind of ideas and feelings that try to suck us down to the bottom of some pit where we struggle with frustration, anger, or discouragement.

Don't get carried away, reader! Try again to say what you think, what you want, or what you feel. If ordinary words aren't enough to express yourself, be creative and find other ways to express yourself: actions that show how you feel; attitudes that show how you think; gestures of humility, love, compassion, consideration, and the like can be excellent ways to say what you want to say. If you are convinced that what you have to say is necessary, uplifting, and true, don't give up! Be a blessing at all times and in all places! The world is hungry for such men.

"Let your speech always be with grace, as though seasoned with salt, so that you will know how you should respond to each person." (Colossians 4:6) – The apostle Paul was an eloquent man, both in his words and in

his deeds.

The Lord Jesus Christ, the Living Word of God, did not always express Himself in words. Take this example:

"So when He had washed their feet, and taken His garments and reclined at the table again, He said to them, "Do you know what I have done to you?" (John 13:12) – Sometimes we must "speak" through acts of great humility in order to be heard.

He had already told his disciples at various times that they should love and serve one another, that the greatest among them should take the position of servant of all. He had already taught them many profound truths using images from nature, such as the parable of the sower, and all the other parables He told the people. He had given unequivocal proof that He was the Messiah, the God–created man who took upon Himself the sins of mankind. He did this through miracles and wonders – when He healed all kinds of diseases, both physical and spiritual; when He walked on the sea to meet the disciples in the boat; when He rebuked the waves and the wind, causing a great calm; when He filled Peter's nets with fish in broad daylight; when He forgave sins; when He mingled with the outcasts of society and lived with them, as in the case of the tax collectors, among others. Yes, the Lord communicated not only with words, but also with the powerful deeds he practiced.

If we follow His way of doing things, we will reap the corresponding rewards.

A moment alone with God to bring down towers

Lord, my God,

Right now I'm looking at the road I've been traveling and I'm looking inside myself. Not everything I see seems right or good. I know that you see me through Christ's perfection, through the exchange He made on the cross for me, but I admit that I often feel as the apostle Paul said: *"For I know that nothing good dwells in me, that is, in my flesh; for the willing is present in me, but the doing of the good is not."* (Romans 7:18)

I have built "towers" in my heart and have forgotten to always honor you in word and deed. I ask your forgiveness. I have sought personal fulfillment in various ways and I recognize that not all of them have been the best. Break down in me, Lord, all the towers that do not glorify You!

Help me, Lord, to understand that I am more than the job I have, the car I drive, or the house I live in. I am much more than my own name, more than my success or failure, more than the natural talents and spiritual gifts you have given me: I am your child, bought at the highest and most excellent price, so I must glorify you with my life, my words and my actions, as it is written: *"For you have been bought with a price: therefore glorify God in your body."* (1 Corinthians 6:20)

Make me a clean temple for the Holy Spirit to dwell in and free me from being a ziggurat in my own name to celebrate myself. I pray that Christ may grow in me and that I may decrease. I pray in the name of the Lord Jesus Christ. Amen.

Chapter 3
The Honest Restorers of the Temple

- Historical data -

Dear reader, let's think together for a moment:

- What would the world be without enterprising men, builders, brave and honest workers?
- What would our cities, towns, villages, and churches be like if there were no one to watch out for erosion and the ravages of time, to rebuild, restore, and preserve buildings?
- What would life be like without people full of God's power who give their all every day, sacrificing all they are and all they have for the sake of others?

Men like this are everywhere in society these days. We've become so accustomed to them that we forget to recognize their work and appreciate them as they deserve!

In this chapter we will talk about the honest restorers of the Temple of Solomon who lived during the reign of Jehoash king of Judah. Below is an outline of some of the pertinent facts of the time. Please follow the arrows:

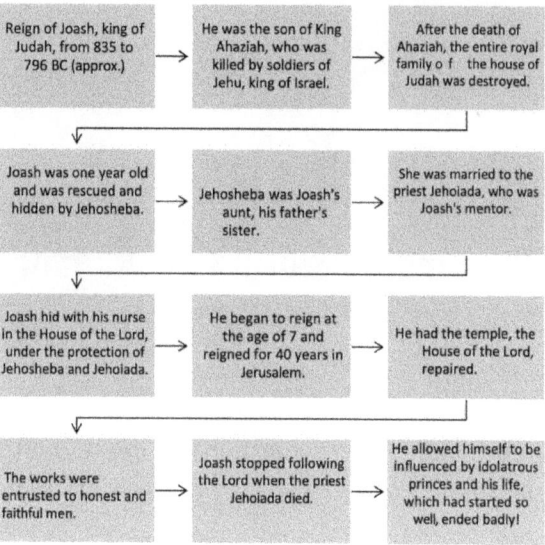

Reign of Joash, king of Judah, from 835 to 796 BC (approx.) →	He was the son of King Ahaziah, who was killed by soldiers of Jehu, king of Israel. →	After the death of Ahaziah, the entire royal family o f the house of Judah was destroyed.
Joash was one year old and was rescued and hidden by Jehosheba. →	Jehosheba was Joash's aunt, his father's sister. →	She was married to the priest Jehoiada, who was Joash's mentor.
Joash hid with his nurse in the House of the Lord, under the protection of Jehosheba and Jehoiada. →	He began to reign at the age of 7 and reigned for 40 years in Jerusalem. →	He had the temple, the House of the Lord, repaired.
The works were entrusted to honest and faithful men. →	Joash stopped following the Lord when the priest Jehoiada died. →	He allowed himself to be influenced by idolatrous princes and his life, which had started so well, ended badly!

The little king

Joash lost his parents when he was one year old. He didn't have time to get to know them well or to keep memories of them. He was saved from the extermination of the royal family of Judah by Jehosheba, the sister of his father, King Ahaziah.

This destruction was ordered by Athaliah, his paternal grandmother, when she learned of her son's death. Athaliah was the daughter of Ahab and Jezebel, a pair of idolatrous kings. Under Jezebel's influence, Ahab established the cult of Baal in Israel. During his 22–year reign, the king sought alliances with Judah to end the constant hostilities between the two kingdoms. The marriage of his daughter Athaliah to Jehoram, king of Judah, was part of one of these alliances. The intention was to unite the two dynasties.

After the death of her parents and many other members of the royal family of Israel, Athaliah, in retaliation, ordered the death of all the descendants of the royal house of David, including Joash, her grandson. The Lord would never allow this to happen, for the Messiah would come from the descendants of David, from the tribe of Judah. So He touched Jehosheba's heart, filled her with courage and used her to save little Joash!

Rescued by Jehosheba, the prince lived hidden in the house of the Lord with his nurse until he was seven years old. The priest Jehoiada, Jehosheba's husband, a godly man, was his spiritual mentor, a kind of surrogate father, master, and counselor.

Meanwhile, Athaliah reigned as regent in Judah.

When Joash was seven years old, Jehoiada arranged for the new king to be crowned safely in the House of the Lord, under the protection of the royal guard. The sounds of joy at the ceremony – clapping, music, and cheers – filled Athaliah's ears. Outraged, she entered the house of the Lord and saw the new king! How could he have escaped? How could he have survived? Treason!" – she shouted. But it was too late to change the course of events. The guards escorted her out of the temple through the ranks and ended her life on the way to the king's house.

Let's pause here for a moment to consider some important issues:

- *The role of a father:* A father's role in a child's life goes far beyond raising and providing for him. When the father, for whatever reason, is no longer present, there is a huge void in the child's life, unless it's an abusive, violent father, dependent on addictions or sick with pathologies that make him unstable and endanger the integrity of the child.

- *Self–evaluation:* Fathers (men) often have a less than positive view of themselves and try to take a secondary role in their child's life. Don't be afraid to be a father! Be a father after God's own heart, taking the Lord as an example – He is a father who understands, loves without conditions, protects, cares, does not abandon, does not accuse or blame, and above all, is present in the life of each of His children!

- *Discovery and appreciation:* Parent–Reader, discover your child's interests, abilities, and potential. Help him develop into a balanced person in all areas – physical, moral and spiritual. The rest will follow.

- *Know how to listen and speak:* It's important to listen to your children when they talk about their dreams and personal projects! It's just as important to know how to give advice without destroying those dreams and plans! We don't always see clearly what God has planned for our children, so there are times in life when it is best to let the Lord speak to them and show them the way, so that we don't find ourselves "fighting God" over their careers or other aspects of their lives.

- *What to avoid:* Avoid telling your child that he never gets anything right or that he always gets everything wrong; avoid telling him that he knows nothing, that he is worthless, that he will always be a failure, that he will never succeed, and so on. First, because it's not true – no one is always wrong and no one "never does anything right"; second, because Christ gave His life for each of our children, just as He did for each of us, so we should not disregard what God Himself considers more valuable than the riches of the whole world; Third, because such words crush everyone's fragile ego – even more so when they come from the mouth of their own father! Children believe adults. If the father (or mother) says their child is no good, the child thinks they must be right. Fourth, because by doing so, without knowing it, you are cursing your child's life instead of blessing it! That's certainly not what you want!

- *Repentance:* If you have ever acted in any of the ways described above, or done anything similar that has deeply hurt or marked your child, this is a good time to talk to God about it, to repent. Don't be afraid to apologize – it only makes you greater in your child's eyes, not less!

- *A life of their own:* Each child will live his or her own life, make his or her own choices, and deal with the good or bad consequences that come his or her way. It is still true that "whatever a man sows, this he will also reap" (Galatians 6:7). Don't expect your child to live your dreams. He has his own dreams. Let him be himself! Be a father, not a policeman! Praise more and criticize less! You can't imagine what praise can do for your child...

The biblical narrative

"Joash was seven years old when he became king, and he reigned forty years in Jerusalem; and his mother's name was Zibiah from Beersheba. Joash did what was right in the sight of the LORD all the days of Jehoiada the priest.

Now it came about after this that Joash decided to restore the house of the LORD. He gathered the priests and Levites and said to them, "Go out to the cities of Judah and collect money from all Israel to repair the house of your God annually, and you shall do the matter quickly." But the Levites did not act quickly. So the king summoned Jehoiada the chief priest and said to him, "Why have you not required the Levites to bring in from Judah and from Jerusalem the levy fixed by Moses the servant of the LORD on the congregation of Israel for the tent of the testimony?" For the sons of the wicked Athaliah had broken into the house of God and even used the holy things of the house of the LORD for the Baals.

So the king commanded, and they made a chest and set it outside by the gate of the house of the LORD. They made a proclamation in Judah and Jerusalem to bring to the LORD the levy fixed by Moses the servant of God on Israel in the wilderness. All the officers and all the people rejoiced and brought in their levies and dropped them into the chest until they had finished. It came about whenever the chest was brought in to the king's officer by the Levites, and when they saw that there was much money, then the king's scribe and the chief priest's officer would come, empty the chest, take it, and return it to its place. Thus they did daily and collected much money. The king and Jehoiada gave it to those who did the work of the service of the house of the LORD; and they hired masons and carpenters to restore the house of the LORD, and also workers in iron and bronze to repair the house of the LORD. So the workmen labored, and the repair work progressed in their hands, and they restored the house of God according to its specifications and strengthened it. When they had finished, they brought the rest of the money before the king and Jehoiada; and it was made into utensils for the house of the LORD, utensils for the service and the burnt offering, and pans and utensils of gold and silver. And they offered burnt offerings in the house of the LORD continually all the days of Jehoiada. (2 Chronicles 24:1–2, 4–14)

"Moreover, they did not require an accounting from the men into whose hand they gave the money to pay to those who did the work, for they dealt faithfully." (2 Kings 12:15)

Slow on the job

After Jehoash married and fathered several children, he made it his mission to restore the Temple of Solomon, the House of the Lord, which was in a poor state of repair. For years, the worship of God had not been a priority for the various kings of Judah, and as a result, the temple had not been maintained and was deteriorating with each passing year.

Since Jehoash had been hidden and raised in the temple itself, he knew its condition well. Restoration of the house of the Lord was urgent! As he was a young man who feared God and wanted to honor Him, he gave orders for the repairs to begin and to be financed by the taxes paid by the people and by voluntary offerings. This money would be collected by the priests and Levites appointed for this purpose.

"He gathered the priests and Levites and said to them, "Go out to the cities of Judah and collect money from all Israel to repair the house of your God annually, and you shall do the matter quickly." But the Levites did not act quickly." (2 Chronicles 24:5)

"But the Levites did not act quickly."

How can we explain this attitude of the Levites? Why didn't they obey the king?

- Because he was too young?
- Because they thought he was inexperienced?
- Because they had a different opinion?
- Because they wanted to use the money for other works in the community instead of the temple?
- Because they didn't care about the decaying state of the House of God?
- Because they didn't want to go out to the cities of Judah and collect the money every year?
- Because they were afraid to take on that responsibility?
- Because they had become accustomed to the degradation of the House of the Lord and were too lazy to try to change the situation?
- Because they didn't love the House of God enough to want to restore it?
- Because they thought that, no matter how hard they tried, they would never be able to replicate the original glory of the temple built by Solomon?

We can't be sure of their motives, the Bible doesn't tell us, but we do know that the Lord's work, whether it be the construction of a building, improvements, repairs, or spiritual work, must be done diligently and promptly, without hesitation or delay, as soon as the means to carry it out have been gathered. When the King commands, we must carry out His orders without hesitation!

The apostle Paul, writing to the believers in Rome, recommended the following:

"Do not be slothful* in zeal, be fervent in spirit, serve the Lord." (Romans 12:11)

About Timothy, the same apostle says these words:

"But I hope in the Lord Jesus to send Timothy to you shortly, so that

---

* Some translations of the Bible use the word "lazy", "remiss", i.e. taking their time to do the work.

I also may be encouraged when I learn of your condition. For I have no one else of kindred spirit who will genuinely be concerned for your welfare. For they all seek after their own interests, not those of Christ Jesus." (Philippians 2:19–21)

May the Lord allow each one of us to dedicate ourselves to God's work, seeking first what is Christ's and leaving personal business in second place!

Emergency meeting

"Then King Jehoash called for Jehoiada the priest, and for the other priests and said to them, 'Why do you not repair the damages of the house?'"

The king went into "zero tolerance" mode! His love for the Lord and the House of the Lord, together with the indolence and inertia of the priests and Levites, led him to call an emergency meeting. At that meeting, several measures were taken, including the following: *"Now therefore take no more money from your acquaintances, but pay it for the damages of the house." (2 Kings 12:7)*

The order couldn't have been more direct. Despite this, we read: *"So the priests agreed that they would take no more money from the people, nor repair the damages of the house." (2 Kings 12:8)*.

Then the wise and elderly Jehoiada took the initiative to place an offering box at the foot of the altar. In it the priests deposited all the money that was brought into the house of the Lord, and the amount grew until it was sufficient to begin the restoration work.

"When they saw that there was much money in the chest, the king's scribe and the high priest came up and tied it in bags and counted the money which was found in the house of the LORD. They gave the money which was weighed out into the hands of those who did the work, who had the oversight of the house of the LORD; and they paid it out to the carpenters and the builders who worked on the house of the LORD; and to the masons and the stonecutters, and for buying timber and hewn stone to repair the damages to the house of the LORD, and

for all that was laid out for the house to repair it." (2 Kings 12:10–12)

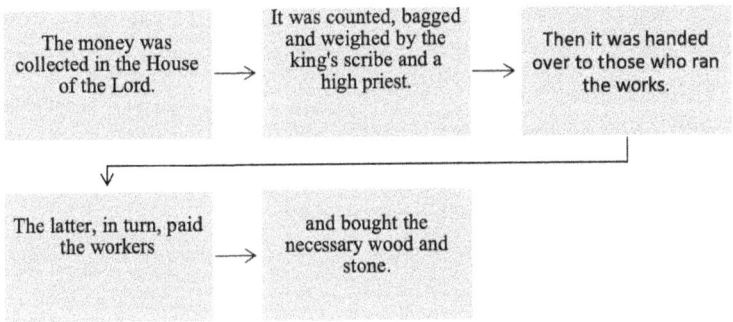

The honest builders

Honesty is the quality of a person who is honest, honorable, and just. An honest man is a precious stone in a world of dishonesty and deceit. If you've ever had to make a deal–buying a car, a house, or any kind of used equipment, you've experienced what it's like to be cheated, or to have someone try to cheat you.

Buying a used car is particularly interesting in this respect. On the one hand, the seller talks about the car as if it were a Rolls Royce or a rare princely carriage, a real finding, a bargain! He assures us that the car is "immaculate", the mileage on the odometer is real, it has always been kept in a garage, there is "no trace of rust", the upholstery looks like it left the factory yesterday, it has never been in an accident, the brakes work perfectly, the transmission is excellent, the headlight bulbs are as good as new, and the engine oil has been changed!

On the other hand, the buyer also puts forward his arguments: he says he 's not sure it's really the kind of car he's looking for; he tries to show an apparent lack of interest; he mentions "in passing" that he's already looked at other options of the same make and the prices were more affordable; he remembers that his wife doesn't like the color... But if the salesman could "make a litle discount", he might even keep the car... In other words, he wants to make a profitable deal, of course! As the Word of God says:

""Bad, bad," says the buyer,

But when he goes his way, then he boasts." (Proverbs 20:14)

For many, the business world is a two-way street: you can fool and you can be fooled. But for an honest man, it is all about integrity. When he is selling, he cannot cheat the buyer, when he is buying, he cannot try to drive down the price to the point where he is not paying the right price for the product.

Honesty is a way to give a good Christian witness. If we decide that even if everyone else is dishonest, we will remain honest because of the fear of the Lord, it will be noticed. If we lose our wallet and it is returned to us by a stranger with all the money and documents it had before, we know we are dealing with an honest person.

In the first letter of the apostle Peter (3:2) we read that the good testimony, *the honest, fearful behavior* of a Christian wife can lead her husband in the direction of God. We're talking about a wife who keeps her husband informed about what's going on in the family, without omitting important information or using deception, who lives honorably and fears God, her Lord. Every husband pays attention to such a woman, in whom he can trust everything.

The apostle Paul speaks of his concern to maintain his own honesty and that of his missionary team wherever they went. They were honest in word and deed, in the administration of the offerings, and in all matters for which they had to give an account to their brothers and sisters in the faith.

"for we have regard for what is honorable, not only in the sight of the Lord, but also in the sight of men." (2 Corinthians 8:21)

In the case we deal with in this chapter, it was essential that all the men who handled the money that went into the temple treasury be honest. This money would pass through several hands before it reached its final destination:

- The hands of the priests who collected it;
- The hands of the scribe and the high priest who counted and weighed it;

- The hands of the foremen;
- And finally, the hands of the workers and the sellers of construction materials.

If there were a weak link in this chain, a pair of dishonest hands, the damage to the work would be great, and the people would lose confidence in the treasurers or the builders. The Scriptures record that there was no failure in the financing and fair payment of the entire work. The men who did it were so honest that there was no need to hold them accountable! They were completely trustworthy. They proceeded correctly in all their dealings and in all payments that were due. We read about them here:

"When they had finished, they brought the rest of the money before the king and Jehoiada; and it was made into utensils for the house of the LORD, utensils for the service and the burnt offering, and pans and utensils of gold and silver." (2 Chronicles 24:14)

They were honest enough to give back to the king all the money left over from the work so that he could use it to make utensils for the various ministries in the house of the Lord! They didn't "embezzle" a single coin. They didn't lie about their expenses. They were honest and anonymous, but honest! The Lord was honored by their honesty. Their names and their conduct will be remembered before God, and none of them will be left without a reward for their honesty.

A moment alone with God

"But seek first his kingdom and his righteousness, and all these things shall be added to you." (Matthew 6:33)

Dear Lord,

Help me to be honest in my words and actions, at home, at work, in business, in everyday buying and selling, in church, even in my prayers!

Help me not to try to appear to be someone other than who I really am. Lord, may You see in me what You saw in Nathanael a man "in whom there is no deceit," a man of complete integrity, an honest man (John 1:47).

Help me to be honest and true, even when it's difficult.

Help me to seek Your kingdom first, confident that all things will be added to me. Then I won't covet what isn't mine, and I won't lay my hand on what isn't mine.

Help me to worship and serve You with an honest heart, not so that others can see or know, but so that You can see and know who I am when no one is watching but You.

In Jesus' name I pray.

Part II

Fearless men

Chapter 1
300 warriors without weapons

- Historical data -

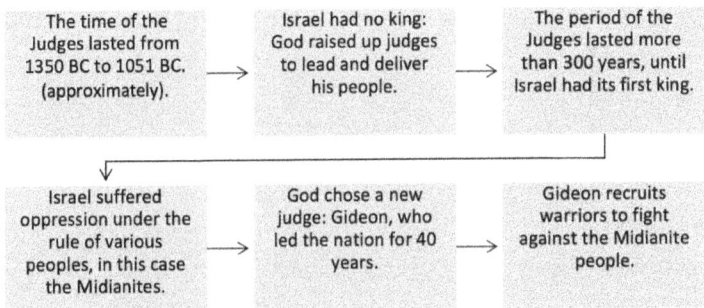

The time of the Judges lasted from 1350 BC to 1051 BC. (approximately). → Israel had no king: God raised up judges to lead and deliver his people. → The period of the Judges lasted more than 300 years, until Israel had its first king.

Israel suffered oppression under the rule of various peoples, in this case the Midianites. → God chose a new judge: Gideon, who led the nation for 40 years. → Gideon recruits warriors to fight against the Midianite people.

Dear Reader,

As you can see from the outline above, in this chapter we will look at the life of Gideon as the leader and judge of Israel and the army of 300 brave soldiers he led into a great battle.

Historians have a hard time pinpointing the exact dates of all these events, but they agree on the veracity of the facts told in Scripture. We will therefore start from what the Word of God tells us and how it tells us that these events took place. It is always our base, the frame that limits the canvas on which all the scenes and all the paintings of each era are depicted.

The descendants of Midian

When we think of the patriarchs mentioned in the Book of Genesis in the Holy Scriptures, namely Abraham, we are led to focus our attention on their extraordinary acts of faith, sometimes forgetting their human side. We forget that these patriarchs, as was said of the prophet Elijah (in James 5:17), were men subject to the feelings common to all men: they suffered when they lost a beloved wife or when they lost a loved one or a son, they felt the same pain and loneliness in the face of life's difficulties that are inherent to being human. For example, after Sarah's death, Abraham married a woman named Keturah. He had six male children by her, including Midian. But Abraham left everything he had to Isaac, the son of God's promise, whom he had by Sarah:

"And I will make you a great nation,
And I will bless you,
And make your name great;
And so you shall be a blessing" (Genesis 12:2)

"And He took him outside and said, "Now look toward the heavens, and count the stars, if you are able to count them." And He said to him, "So shall your descendants be." (Genesis 15:5)

Abraham's other sons, that were many, received gifts and were sent to other countries, away from Isaac and his descendants. God's promise would be fulfilled in Isaac and his descendants, not in any other son of Abraham. Never in Midian!

Abraham didn't favor one son over the other. Rather, he was obeying a higher command! When the Lord commanded, His servant obeyed!

The centuries have passed, and we find two nations facing each other: Israel, the descendants of Isaac and his son Jacob (whom the Lord renamed Israel), and the Midianites, the descendants of Midian, Abraham's son, but outside of God' s special promise to the patriarch.

The Midianites were nomads. The Israelites were sedentary. After being freed from slavery in Egypt and wandering in the wilderness for 40 years (for various reasons that are not part of our subject), they finally entered the Promised Land – Canaan.

However, the land was occupied by other peoples whom the Lord wanted to drive out of the area, so the descendants of Abraham, Isaac, and Jacob had to conquer it. God blessed Israel in the process of conquest, but whenever the people deviated from the basic principles of the Law of Moses, especially regarding idolatry, the Lord called Israel to account and allowed other nations to dominate them for longer or shorter periods of time.

When Israel recognized its sin and sought God, the Lord raised up a judge to deliver the nation. History repeated itself many times...

The time we're referring to in this chapter includes a period of 7 years of oppression under the Midianites that the Lord allowed so that His chosen people would return to Him. For 7 years, the Midianites periodically raided Israel's territory and plundered everything they found. The harvest season, instead of being a time of joy and celebration for Israel, was a time of fear. It meant that the Midianites would soon appear and take everything they found. They would be joined by other peoples from the east, as well as the Amalekites, to carry out the plunder. The people of Israel sought refuge in mountain caves, pits and fortifications. In the book of Judges we read the following:

> *"Then the sons of Israel did what was evil in the sight of the LORD; and the LORD gave them into the hands of Midian seven years. The power of Midian prevailed against Israel. Because of Midian the sons of Israel made for themselves the dens which were in the mountains and the caves and the strongholds. For it was when Israel had sown, that the Midianites would come up with the Amalekites and the sons of the east and go against them. So they would camp against them and destroy the produce of the earth as far as Gaza, and leave no sustenance in Israel as well as no sheep, ox, or donkey. For they would come up with their livestock and their tents, they would come in like locusts for number, both they and their camels were innumerable; and they came*

into the land to devastate it. So Israel was brought very low because
of Midian, and the sons of Israel cried to the LORD." (Judges 6:1–6)

The Angel of the Lord meets a man in hiding

"Give ear, O LORD, to my prayer;
And give heed to the voice of my supplications!" (Psalm 86:6)
"The sacrifices of God are a broken spirit;
A broken and a contrite heart, O God, You will not despise." (Psalm 51:17)

The Lord heard every prayer and every petition. He saw the suffering of the people for seven years. At some point, He noticed their repentance and the humility with which they interceded for the nation. And God moved again on behalf of Israel!

Reader, it's never too late to turn to God with a broken and repentant heart, whatever your story. Come back again, and let God act on your behalf! "(...) I act and who can reverse it?" (Isaiah 43:13)

For a moment, imagine that God says something like this to you:

If I decide to reach out and set you free, who's to say I can't?

Who will stand in my way, who will keep me from embracing you, from staying close to protect you?

And if I want to bless you beyond measure, let the windows of heaven open wide and streams of gentle blessings descend, fall upon you, envelop your life,

Who will tell me what I can and cannot do?

If I want to lift you up again, carry you, heal your every wound and shelter you under My wings, who will be proud or haughty enough to oppose Me, the Eternal God?

Receive, son, my fatherly love! Rest here in my strong embrace.

What you cannot do, let me it for you! I am your shepherd, you will lack nothing.

When I act, no one will stop me!

A man named Gideon, son of Jehoash, was threshing some wheat... in a winepress! Not on the threshing floor, in the sun and wind, so that the mill would be blown away and the grain would be clean, but in a winepress. Closed. Hidden. All because of the Midianites who were camped nearby, waiting for the harvest to be done, the grain to be clean, the fruit to be ripe, and then they would come like birds of prey to pounce on their prey!

The Angel of the Lord† watched this man in hiding. From time to time he would lift his eyes from his work and scan the nearby horizon: no one in sight. He could go on threshing his wheat.

Gideon was under the watchful eye of the Lord, but he didn't know it until the Lord became visible and spoke to him:

> *Then the angel of the LORD came and sat under the oak that was in Ophrah, which belonged to Joash the Abiezrite as his son Gideon was beating out wheat in the wine press in order to save it from the Midianites. The angel of the LORD appeared to him and said to him, "The LORD is with you, O valiant warrior." Then Gideon said to him, "O my lord, if the LORD is with us, why then has all this happened to us? And where are all His miracles which our fathers told us about, saying, 'Did not the LORD bring us up from Egypt?' But now the LORD has abandoned us and given us into the hand of Midian." The LORD looked at him and said, "Go in this your strength and deliver Israel from the hand of Midian. Have I not sent you?" He said to Him, "O Lord, how shall I deliver Israel? Behold, my family is the least in Manasseh, and I am the youngest in my father's house." But the LORD said to him, "Surely I will be with you, and you shall defeat Midian as one man." (Judges 6:11–16)*

From the passage we just read, let's highlight the expressions that reveal what the Lord thought of Gideon:

- *"valiant warrior"* was not Gideon's idea of himself;
- *"Go in this your strength"* – What strength? He was hiding, threshing his precious wheat, and then he was going to hide it as best as he could so the Midianites wouldn't steal it from him!

---

† We think it is the Lord Himself, since He speaks as God, with God's attributes and authority.

- *"and deliver Israel"* – He could barely save himself and his possessions, let alone a nation!

Here is proof that what we think of ourselves is not always what the Lord thinks! The Lord looks at the raw clay and sees the final piece He will make out of that clay. We see only the clay, nothing but the clay.

"But the LORD said to Samuel, "Do not look at his appearance or at the height of his stature, because I have rejected him; for God sees not as man sees, for man looks at the outward appearance, but the LORD looks at the heart." (1 Samuel 16:7)

The Lord looked into the heart of the man hidden in the wine press and saw the warrior and leader who would lead His chosen people to great victory!

Reader, perhaps there is a warrior or leader hiding in your heart, just waiting for the right moment to emerge! Perhaps there is a man of priceless integrity and worth, but too insecure to stand up and fight for what is just and right! Perhaps there is a man with the faith to move mountains, but who has not yet had the courage to exercise the authority that comes from the Lord over those mountains! The eyes of the Lord are upon you. He is near and has a mission to give you!

Gideon's answer to the angel of the Lord was like the sigh of a weary man, weary of many things:

- Of waiting for God's deliverance;
- To hide their livelihoods for themselves and their families, year after year;
- Fearing the next onslaught of the Midianites;
- Fear of being deprived of the property to which God had promised him in the land of milk and honey;
- Fear of not having enough to feed his family for the rest of the year;
- To pray without receiving the desired answers;
- To see his people follow the pagan customs of the peoples of the land of Canaan! Even his own father had an altar to Baal, the chief "god" of the Canaanites! How low one had to sink to

reach this deep pit of rebellion against the Lord! How far from God was Jehoash, his beloved father!

All these things tired the man to whom the angel of the Lord spoke. How much longer would it take for the people to learn the urgent need to abandon these evil ways and return to the Lord, their God, the Creator of heaven and earth? No more time is needed, Gideon! It's time! And you are the instrument that God will use to deliver His people!

"Now on the same night the LORD said to him, "Take your father's bull and a second bull seven years old, and pull down the altar of Baal which belongs to your father, and cut down the Asherah that is beside it; and build an altar to the LORD your God on the top of this stronghold in an orderly manner, and take a second bull and offer a burnt offering with the wood of the Asherah which you shall cut down." (Judges 6:25–26)

Did he understand correctly? To tear down his father's idolatrous altar? To offer as a burnt offering an ox that his father had kept and hidden from the midianites during the seven years of successive attacks?

Yes, that's right! If something needs to be straightened out, it should start at home...

Reader, maybe it's time to change something in or around you. Maybe it's time to stop keeping your faith private and hidden, to come out and talk about it in broad daylight, in the sun and the wind, taking advantage of every opportunity the Lord gives you, taking an unassailable position because you are absolutely certain of who the Lord your God is!

A great success

Gideon needed a large army to face the huge crowd of Midianites! In a short time, he was able to gather 32,000 men! It was an interesting number, but when you look at the hundreds of thousands of men in the opposing army, what were 32,000? Nevertheless, he led his men to the place where they could set up camp, at the spring of Harod (which means *trembling*).

He had to organize them and teach them discipline and obedience to his orders, among many other rules essential to the survival of the soldiers. Then some problems arose:

- The Lord commanded all who were *timid and afraid* to retreat! Gideon watched in horror as hundreds and then thousands of men retreated!

- The report of the new count of those remaining said that 22,000 soldiers had retreated! There were only 10,000 left. How was that possible? *"What can I do with 10,000 men?"* You, Gideon, can do nothing, but God can do whatever is necessary!

- Then came the Lord's second command: *"There are still too many people..."* No, Lord, no, please! They are not too many! They are few compared to the Midianites! *"Bring them down to the water, and there I will test them; whom I say to you, 'This one will go with you,' that one will go with you..."* Mercy, Lord!

- Out of the 10,000 men, the Lord chose 300 and specified the weapons they were to take into battle, which Gideon distributed to each soldier in his meager army: a trumpet and an empty jar that would hold a burning torch on the day of battle.

The army was ready and waiting! Did the men exchange looks of doubt? Probably not! The 22,000 timid and fearful men would have done so, but not these 300! They would have worried about the thirst or hunger they might feel during the battle, where to find water, where to rest, where to hide. No, not those 300! That would be for the 9,700 who were running around quenching their thirst, oblivious to the dangers around them, focused on themselves, dominated by the desire to get enough water, without enough self–discipline to restrain themselves and keep watch at all times!

The men were divided into three companies of 100 men each. They were positioned in strategic locations and told to wait for Gideon's signal!

What an awesome test of faith! What extraordinary men these 300 were! They were willing to die for their people if they had to.

If they had fallen on the battlefield, we wouldn't even know their names to honor them properly! They would all be like "unknown soldiers". 300 nameless men, 300 soldiers, 300 of God's anointed laid down their lives to obey not only Gideon's command, but their Supreme Commander, the God of Israel!

On the night of the battle, they were not commanded to run with sword or spear drawn and attack the Midianite camp. Swords and spears were not part of the equipment provided. Rather, they were asked to stand firm at their posts and, at Gideon's command, blow their trumpets, break their pitchers, and lift their burning torches high!

Victory

Victory belongs to the Lord!

"The LORD said to Gideon, "I will deliver you with the 300 men who lapped and will give the Midianites into your hands; so let all the other people go, each man to his home." So the 300 men took the people's provisions and their trumpets into their hands. And Gideon sent all the other men of Israel, each to his tent, but retained the 300 men; and the camp of Midian was below him in the valley." (Judges 7:7–8)

"Now the Midianites and the Amalekites and all the sons of the east were lying in the valley as numerous as locusts; and their camels were without number, as numerous as the sand on the seashore." (Judges 7:12)

Shrouded in the night, each man took his place in silence, all aware of the risks they were taking, but all convinced that the victory would be the Lord's! They didn't choose the place they wanted, or the place that seemed best, safest, most comfortable, or most secure. Each obeyed the orders he had, and each waited quietly for the orders of the leadership.

"When the three companies blew the trumpets and broke the pitchers, they held the torches in their left hands and the trumpets in their right hands for blowing, and cried, "A sword for the LORD and for Gideon!"

Each stood in his place around the camp; and all the army ran, crying out as they fled. When they blew 300 trumpets, the LORD set the sword of one against another even throughout the whole army; and the army fled as far as Beth–shittah toward Zererah, as far as the edge of Abel–meholah, by Tabbath. The men of Israel were summoned from Naphtali and Asher and all Manasseh, and they pursued Midian." (Judges 7:20–23)

"Cease striving and know that I am God;
I will be exalted among the nations, I will be exalted in the earth." (Psalms 46:10)

With one voice, they proclaimed: *"A sword for the LORD and for Gideon!"* Then they smashed the pitchers and lifted the burning torches high into the air while they blew the trumpets!

- *"the three companies blew the trumpets"* – 300 men at once, like a single trumpet, the trumpet of God, sounding loudly, breaking the silence of the night! *"For if the bugle produces an indistinct sound, who will prepare himself for battle?"* (I Corinthians 14:8) The Midianite camp awoke from a deep sleep with the sound of battle, unable to see which side the attacking "army" was coming from!

- *"broke the pitchers"* – Unbroken jars don't reveal the light within. If the Lord doesn't break us, if we don't break ourselves at His feet, the light of Christ won't shine to illuminate anyone's night!

- *"held the torches in their left hands"* – With their arms raised, 300 men raised their burning torches. Burning. Not just smoldering. Not just with traces of having been lit once, but now lit with a strong flame that breaks the darkness! *So be your life and mine, reader!*

There was no rehearsal. There was no second chance if it didn't go well the first time...it would have been the certain end of those 300 men with nerves of steel and precious faith!

There are times in life when we must follow the Lord's specific instructions without another opportunity like this; similar, perhaps, but not the same. These 300 men turned their obedience of faith into a decisive step toward Israel's victory.

Of course, other warriors were called and went into battle. But the victory was already won! Then thousands of brave men pursued and defeated those who had tormented them for 7 years! But it was after... After the terror and chaos that unexpectedly unleashed on the Midianite camp. After the sound of the trumpets, the broken pitchers, and the lifted torches. After.

"[T]he LORD is not restrained to save by many or by few." (1 Samuel 14:6)

A moment alone with God

Dear Lord,

I know that nothing is impossible for You. Even though my opponents seem to be much stronger, more numerous and more cunning than the few friends who are on my side, I know that I can achieve victory through Christ, my Redeemer.

Give me the faith to let you break me, for I am nothing more than a pot of clay in your hands. Give me the strength to hold high the torch lit with the flame of faith and give me the courage to blow the trumpet and speak of Your great love and Your great power!

I pray in the name of the Lord Jesus Christ.

Chapter 2
2600 right-handed men

Dear Reader,

In this chapter we are going to talk about 2,600 extraordinary but anonymous men, highly trained soldiers, who lived in Judah during the reign of Uzziah. Here are some interesting facts about this king:

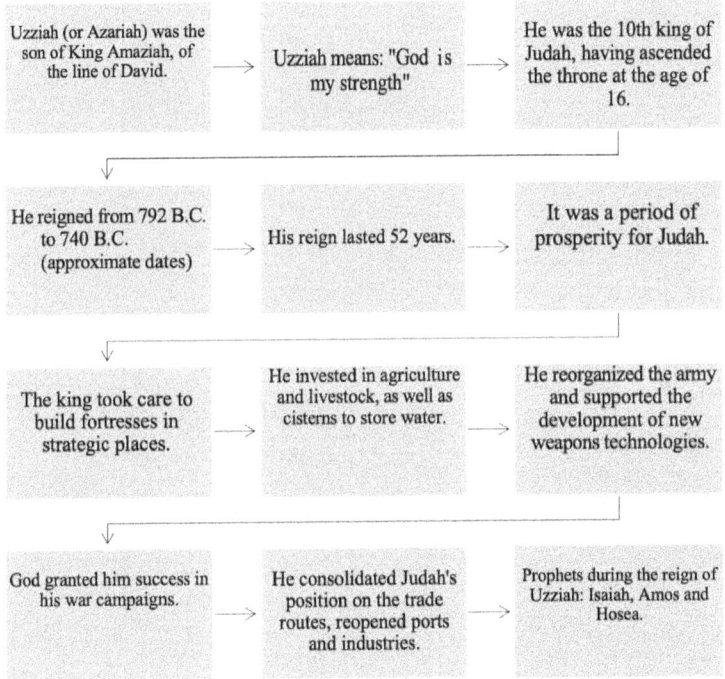

Uzziah (or Azariah) was the son of King Amaziah, of the line of David. → Uzziah means: "God is my strength" → He was the 10th king of Judah, having ascended the throne at the age of 16.

He reigned from 792 B.C. to 740 B.C. (approximate dates) → His reign lasted 52 years. → It was a period of prosperity for Judah.

The king took care to build fortresses in strategic places. → He invested in agriculture and livestock, as well as cisterns to store water. → He reorganized the army and supported the development of new weapons technologies.

God granted him success in his war campaigns. → He consolidated Judah's position on the trade routes, reopened ports and industries. → Prophets during the reign of Uzziah: Isaiah, Amos and Hosea.

Biblical narrative

The biblical text that will serve as the basis for this study is found in 2 Chronicles:

> "Moreover, Uzziah built towers in Jerusalem at the Corner Gate and at the Valley Gate and at the corner buttress and fortified them. He built towers in the wilderness and hewed many cisterns, for he had much livestock, both in the lowland and in the plain. He also had plowmen and vinedressers in the hill country and the fertile fields, for he loved the soil. Moreover, Uzziah had an army ready for battle, which entered combat by divisions according to the number of their muster, prepared by Jeiel the scribe and Maaseiah the official, under the direction of Hananiah, one of the king's officers. The total number of the heads of the households, of valiant warriors, was 2,600. Under their direction was an elite army of 307,500, who could wage war with great power, to help the king against the enemy. Moreover, Uzziah prepared for all the army shields, spears, helmets, body armor, bows and sling stones. In Jerusalem he made engines of war invented by skillful men to be on the towers and on the corners for the purpose of shooting arrows and great stones. Hence his fame spread afar, for he was marvelously helped until he was strong." (2 Chronicles 26:9–15)

Advice and strategies for success

King Uzziah was 16 years old when he was placed on the throne of Judah by his own people in place of his father, Amaziah. The Scriptures also record the name of his mother, Jecoliah of Jerusalem. This name means: *"God has prevailed"* or *"God is able"*. We know that in the Bible, a person's first name was important because of its meaning and because it usually reflected that person's character or particular experience with God. It is likely that Jecoliah had experienced firsthand how the Lord prevails and overcomes, how He is able to deliver, sustain, and care for those who love Him even in the most adverse circumstances.

The young king Uzziah knew how to seek the Lord's guidance – "He continued to seek God" – and to listen to the counsel of Zacharias, a man wise in the visions of God, as recorded in 2 Chronicles 26:4–5.

Reader, if you're still a young person, this is the perfect time to seek God with all your heart, follow the practical teachings of the Scriptures, and make choices and decisions under the guidance of the Lord. By doing so, you will have a life in front of you according to what God has planned for you. What's more, in a few years you will not regret that you did not give everything that concerns you to the Lord.

"Remember also your Creator in the days of your youth, before the evil days come and the years draw near when you will say, "I have no delight in them" said the wise Solomon in Ecclesiastes 12:1.

Heed the advice of your elders, listen to their stories, discover the wisdom in them, and understand their physical weaknesses.

One day, the frailty that comes with age may knock on your door. As you continue to read Ecclesiastes chapter 12, you will see that strength and youth are fleeting, even if they don't seem so now, at the peak of your strength and vigor.

On the other hand, if you are older and already mentally counting down the years ("How many years do I have left?"), then this is the "perfect" time to seek God's presence, fix everything that isn't right, make amends for any mistakes that can be amended, live each day to the fullest, surround yourself with the people you love, hug, cultivate healthy habits, and laugh, laugh, and be very happy! Worry more about leaving your family a good emotional and spiritual legacy than a material one, which will quickly run out when you leave.

Leave them fantastic memories! Give them a good family name, an honorable name, an example of character! Become the special person they want to spend time with. Don't skimp on praise and recognition of their good work and efforts! Listen to them and encourage them! Next comes eternity with God, the supreme ecstasy of contemplating the Lord, endless happiness! What an extraordinary hope! But you will no longer be able to do anything for your friends or your family! It will be too late... Now is the right time, the most perfect time!

"Yet you do not know what your life will be like tomorrow. You are just a vapor that appears for a little while and then vanishes away." (James 4:14)

But let's say you're well into adulthood, healthy and very active! You've already gained a lot of experience and developed a good level of maturity, you've learned a lot from some of the mistakes you've made and also from observing the mistakes of others. you have found financial and emotional stability. I must tell you that this is an excellent and urgent time to seek God with all your heart, with all your strength and with all your soul!

If you haven't found the right pace in your walk with God, you can start over now and adjust to the pace the Lord has for you.

Take a quick look back and see how God has stood by you and been faithful to you! Then look ahead again and decide to walk with God from now on, so that you won't have anything important to regret in the future.

In just a few verses, the Bible tells the story of a man named Enoch. Look, reader:

"Then Enoch walked with God three hundred years after he became the father of Methuselah, and he had other sons and daughters. So all the days of Enoch were three hundred and sixty–five years. Enoch walked with God; and he was not, for God took him." (Genesis 5:22–24)

"By faith Enoch was taken up so that he would not see death; AND HE WAS NOT FOUND BECAUSE GOD TOOK HIM UP; for he obtained the witness that before his being taken up he was pleasing to God." (Hebrews 11:5, emphasis added)

For this man, Enoch, the birth of a son was such an important milestone that it changed the course of the rest of his life: he began to walk in fellowship with God until he entered His presence forever!

Reader, don't take this lightly. Be blessed and be a blessing to everyone in your household, your work environment and your daily contacts! Don't forget your friends near and far. Don't forget to use all your knowledge and all the other means with which the Lord has blessed you to make the world a better, fairer, cleaner, and safer place for the present generations and all those who come after you. Continue to build a unique legacy to leave to those who deal with you daily, who will remember you on *countless occasions, long after you have retired or moved to the "Upper Home". Become the person every friend wants to meet for coffee, a walk, or just a chat.*

There is no magic formula for success. We need a lot of commitment and a lot of dependence on the Lord! It's all very hard work, demanding, tough, limited in time and space. The hours of concentration and effort required of us are immense! If there were a mathematical formula for true success, it would certainly include God as the main factor. So let's not exclude Him in any way!

I challenged a young mathematician to come up with a formula for success in the Christian life that any of us could apply without having in–depth knowledge of the subject. Here is her valuable contribution:

The formula
Special contribution by Beatriz Leite

Whether we are fans of mathematics or not, it is undeniable that it is fundamental to humanity. It provides order and prevents chaos in so many areas of our lives, such as managing our money, using a map or GPS, watching sports and interpreting their statistics, etc.

It's in the 7th grade that we begin to learn about mathematical formulas and functions – expressions, rules or laws that tell us the relationship between two variables (one dependent and one independent). To make the distinction between dependent and independent variables clear to the reader, let's look at the following practical example:

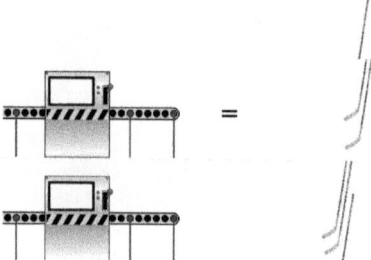

Vemos que o produto final (o que sai da máquina) depende do produto inicial (o que
~~entre na máquina). Ou seja, o número de sticks de háquei em patins que produzimos~~

We can see that the final product (what comes out of the machine) depends on the initial product (what goes into the machine). In other words, the number of roller hockey *sticks* we produce depends on the number of wooden planks we initially have available (and never the other way around). Thus, our independent variable is the number of wooden planks used and the dependent variable is the number of *sticks* that can be produced with the same number of planks.

In addition, we can see that with 2 wooden planks we produce 6 field hockey *sticks* and with 5 wooden planks we produce 15 sticks. From this we can see how the two variables are related and therefore we can conclude that each plank allows us to produce 3 roller hockey *sticks:*

No. of sticks produced = 3 x No. of wooden planks

If we wanted to write this function in a more mathematical way, it would look like this:

$$Sticks\ (t) = 3 \times t$$

The function (called *"Sticks"*) tells us that the number of *sticks* produced depends on the variable in brackets (t), where t is the number of sheets used.

Let's assume that the reader can read 15 pages of this book in one hour. If you wanted to write a function to return this information, it would look like this

Total number of pages read = 15 x number of hours to read Pages (h) = 15 x h

In other words, the Pages function tells us the number of pages the reader has read in a given time interval and depends on the variable h, where h is the number of hours in the same interval.

If the reader has read for 10 hours → Pages (10) = 15 x 10 = 150 pages
If the reader has read for ½ hour → Pages(0.5) = 15 x 0.5 = 7.5 pages

If the reader is faster and can read 20 pages per hour, the function is

Pages (h) = 20 x h

If the reader has read for 10 hours → Pages (10) = 20 x 10 = 200 pages
If the reader has read for ½ hour → Pages (0.5) = 20 x 0.5 = 10 pages

The graph of these functions would look like this:

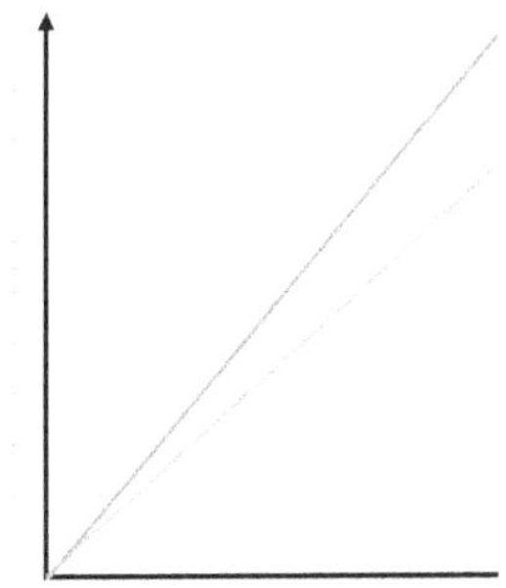

Hours of reading

The yellow line corresponds to the graph of the function Pages (h) = 15 x h, while the green line corresponds to the graph of the function Pages (h) = 20 x h. We can see that the green line is steeper than the yellow line, and this is because the reader of the green line is reading **faster** than the reader of the yellow line. In the first hour (as shown by the dashed line in the graph below), the green line tells us that the reader has read 20 pages and the yellow line tells us that the reader has read only 15 pages. In the second hour, the green line indicates that the reader has read 40 pages and the yellow line indicates that the reader has read only 30 pages.

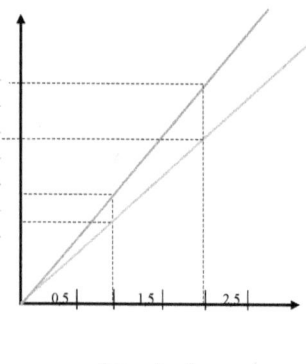

Hours of reading

Thus, the value that precedes the independent variable in the function affects the **slope** of our graph. The higher this value, the steeper the corresponding line.

Knowing all this, we are able to create a formula that shows us the relationship between the variable of time (every second, minute, hour and day of our lives) and the variable of success in the Christian life. In other words, a formula that relates the success of the Christian life to the way we use our time.

Our function can be written as follows

Success (t) = ? * t

The dependent variable is the success of the Christian life, since it depends directly on what we do with our time (t) each day.

The missing value in the function, as shown in the previous example, corresponds to the slope of the line on the graph. We already know that the higher this value, the steeper the line, and vice versa.

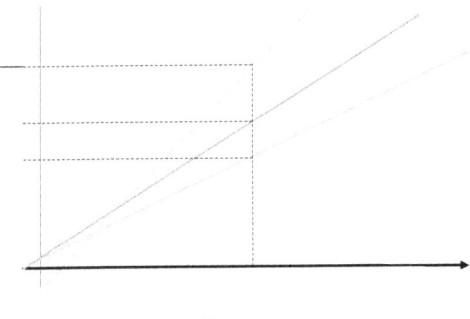

Time

The Christian's goal will be to achieve as much success as possible in the near future. In other words, the Christian's goal is to make the slope of his graph as steep as possible, since this means achieving greater levels of success in shorter intervals of time. In the graph above, we can see that at any given point in time, the level of success achieved by the red line (= c) was greater than the levels of the other lines.

Therefore, our priority is to maximize the slope given by our success function. And what does that slope include? Very important factors like

Faith + Love + Fear of God + Gifts + Talents + Commitment+Work + Humility + Dependence on the Holy Spirit + Obedience (...)

One final note. We might be led to think that those Christians who will live longer than we will necessarily have more opportunities and possibilities to achieve high levels of success in the Christian life. However, such a conclusion is not necessarily correct. What we should be concerned about is not the amount of time God gives us on earth, but rather the slope we put on our success function, i.e., how we make the most of the time God gives us!

In the following graph, we see that the Christian who had less time on earth (represented by the light blue line) achieved a higher level of success than the Christian who lived for more years. This is because the slope of their success function was high.

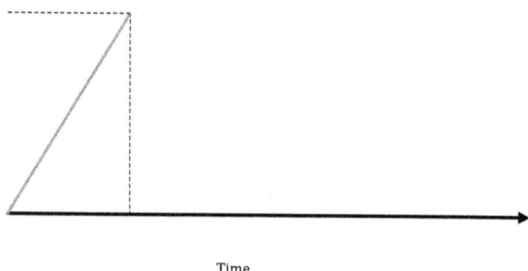

Time

2600 right-handed men

In addition to the various reforms and initiatives King Uzziah introduced during his reign, let's highlight the army. The NVT (New Transformation Version) of the Bible describes Uzziah's army this way (2 Chronicles 26:11–14):

> "Moreover, Uzziah had an army ready for battle, which entered combat by divisions according to the number of their muster, prepared by Jeiel the scribe and Maaseiah the official, under the direction of Hananiah, one of the king's officers. The total number of the heads of the households, of valiant warriors, was 2,600. Under their direction was an elite army of 307,500, who could wage war with great power, to help the king against the enemy. Moreover, Uzziah prepared for all the army shields, spears, helmets, body armor, bows and sling stones."

I - THE MEN OF THE ARMY:

a. *"valiant warriors"*‡ – the recruitment and training given to these

---

‡ In various translations of the Bible the expression "well–trained" is translated by the word *"right–handed"*, which indicates a person endowed with dexterity or agility in handling instruments (in this case, weapons), someone who is shrewd and perceptive or someone who predominantly

2,600 men was top quality; we need warriors like this in the ranks of the army of faith!

b. *"ready for battle"* – ready: prepared, active, attentive, diligent, available, fearless! They were the first to present themselves.

c. They were the first to go to the front and the last to withdraw after their victory.

d. *"combat by divisions"* – according to each one's specialty in handling weapons;

e. *"prepared"* – they were not an occasional group of brave "guys" to intervene, on a one–off basis, to defend some land or some of the country's interests. They were men with one purpose: to defend their nation and their king to the point of giving their lives, if necessary! They were always summoned and always organized! How we need this in the church of the Lord Jesus Christ!

f. *"under the direction of Hananiah, one of the king's officers"* – their commander was an officer of the king, a man in the king's complete confidence, with whom he spoke regularly, a man who had privileged contact and access to the king, who received direct orders from him and faithfully passed them on to his warriors. O God, give us "officers" of King Jesus, who listen daily to his voice, to his orders, and pass them on faithfully to the other warriors in the Christian army! In the name of Jesus I pray.

II - WAR EQUIPMENT

We are immediately led to think of the armor of God, as it is described in the apostle Paul's letter to the Ephesians, in chapter 6, between verses 11 and 17:

> *"Put on the full armor of God, so that you will be able to stand firm against the schemes of the devil. For our struggle is not against flesh and blood, but against the rulers, against the powers, against the world*

---

uses their right hand in their work.

forces of this darkness, against the spiritual forces of wickedness in the heavenly places. Therefore, take up the full armor of God, so that you will be able to resist in the evil day, and having done everything, to stand firm. Stand firm therefore, having girded your loins with truth, and having put on the breastplate of righteousness, and having shod your feet with the preparation of the gospel of peace; in addition to all, taking up the shield of faith with which you will be able to extinguish all the flaming arrows of the evil one. And take the helmet of salvation, and the sword of the Spirit, which is the word of God."

a. *Shield* – an essential piece for the soldier's protection: If the warrior is right–handed, the shield is used with the left arm; being a mobile piece, the greater the warrior's mastery in its use, the more likely he is to defend himself well and not be injured in battle; The shield must be raised with the intention of preventing any of the opponent's weapons from hitting the warrior, be they swords, spears or arrows (incendiary or not);

b. *Spear* – a throwing weapon designed to hit an opponent (or other target) from a distance, but which can also be used in close combat; it requires precise aim, a steady gaze, arm strength, and determination to be effective;

c. *Helmet* – an essential piece: with it, the warrior protects the head, the center of thought and reason; without it, the soldier is easily neutralized; it is compared to salvation in Ephesians 6, mentioned above.

d. *Breastplate* – protective garment for the vital organs: heart, lungs, and all the organs in the abdomen; in Ephesians 6, this garment is compared to the righteousness the Christian receives through Christ and His perfect sacrifice on the cross of Calvary;

e. *Bow* – a long–range weapon designed to shoot arrows; its handling requires long and disciplined training, excellent visual acuity, arm strength, good balance and determination, because the arrow you shoot doesn't turn back! It's like words that, once spoken, become a blessing or a curse, healing or hurting.

We need to be sure of what we are going to say, whether we should say it or not, with what intention, with what force and in what direction! We can hurt without meaning to, kill with "friendly fire," hurt our own army, hurt the people we love and who love us back.

f. *Sling* – In the tradition of King David, from whom Uzziah descended, this king had his 2,600 chosen warriors equipped with slings. The sling is also a ranged weapon that allows you to attack from a distance. It could be used to throw stones, as in this case, or other objects. There were several types of slings, but King David's was for throwing stones, just like the ones King Uzziah gave to his warriors.

The sword does not appear on this list of weapons. It seems that every warrior already had his own. In Ephesians 6:17, the sword as a basic piece of God's armor is called *"the sword of the Spirit, which is the Word of God."* This spiritual armor is available to every child of God. He or she must put it on daily and remain at his or her post, alert, prepared, and ready to spring into action at any moment.

Each soldier is responsible for taking care of his own armor. This task is not left to the army officer, any more than it is the responsibility of the church pastor!

No soldier leaves his weapons in a corner, hoping that "someone" will clean them, oil them, check that the blade of the sword is sharp, that the bowstring is taut, that the quiver has arrows, that the spear hasn't lost its point, that the shield is firmly clamped, that the shoes are in good condition, as well as the breastplate and the belt... This care of the armor is part of the warrior's daily work. The good condition of each piece of armor contributes to its good performance in battle. Life depends on the condition of the armor, as does victory or defeat in the daily struggles of each Christian.

Very often, the battle is not fought collectively, but is individual and takes place on a large and difficult field: the mind.

Satan, the Christian's greatest adversary, misses no opportunity to attack with his full arsenal! He chooses the weapons that he hopes will be the most effective in defeating each person. He doesn't attack everyone in the same way, with the same weapons or the same strategy. He is observant.

He knows what moves each one of us, what worries us or calms us, what makes us sad or happy, what weakens us or strengthens us, and so on. Therefore, both the weapons and the method, the place and the time are carefully chosen to bring down each child of God, or at least to rob them of some precious possession, such as Holy Communion with the Lord, peace, or family harmony!

He is also preparing collective attacks – against the Church, or whole families, or groups of Christians who are committed to serving God and whose efforts are bearing abundant fruit for the Kingdom of the Lord. Therefore, it is essential – a non–negotiable condition in the career of faith – to remain vigilant, to maintain fellowship with other children of God who are as imperfect as we are, but who carry in their souls the burning desire to fight the good fight of faith.

However, we read of other war machines that were developed and placed in strategic defensive positions during the reign of Uzziah. They were not individual weapons and were not easily transported. Let's look again at the account in 2 Chronicles 26:15 (NVT):

"In Jerusalem he made engines of war invented by skillful men to be on the towers and on the corners for the purpose of shooting arrows and great stones. (...)"

"engines of war invented by skillful men" – There is a translation that uses the term "engineers" instead of "experts. These machines were used to support the army when the battle took place around the walls of Jerusalem. They were machines for shooting arrows and throwing large stones from the towers and the corners of the walls.

I would ask the reader to allow me the liberty of comparing arrows to prayer, even because in the last verses of Ephesians 6, about the armor of God, we read: *"With all prayer and petition pray at all times in the Spirit, and with this in view, be on the alert with all perseverance and petition (...)"*. Let's think of prayer as an arrow shot at the throne of grace! We throw our petitions at the feet of the Lord and wait for the answer to fall on our battlefield and bring us victory.

King Uzziah's war machines also threw great stones, stones that crushed the enemies below, around the walls. Let us pray in the power of the Word, in the power of the Holy Spirit, so that every adversary who rises up against us or against the name of the Lord will be crushed by the Eternal Rock, by the name of the Lord Jesus Christ, the Living Word of God!

More than 307,500 anonymous warriors

"Under their direction was an elite army of 307,500, who could wage war with great power, to help the king against the enemy." (2 Chronicles 26:13)

"Under them were reinforcement troops numbering 307,000, with 500 of them on constant alert—a strong royal defense against any attack" (2 Chronicles 26:13, MSG)

If the reader has carefully read the different translations of the same verse in 2 Chronicles 26:13, they speak of an *elite* army of 307,500 men *"who could wage war with great power"*.

The 2,600 men skilled in warfare had this powerful, elite army of 307,500 men under their command. The preparation of these great warriors depended on the 2,600 officers who commanded them. These men made tough decisions every day. They put some in the front and others in the back. They trained each man for a specific mission. Each man's life depended on the brother in arms at his side, and vice versa.

- Who were the best at archery? – They would belong to a certain unit.
- Who were the best with the stone sling? – They were a different unit.
- Who could best wield the spear from his horse? – The spearmen were an equally powerful phalanx.
- Who operated the war machines? – They had to be highly trained soldiers, specialized in the operation and handling of these machines, elite soldiers of the elite soldiers. The victory

of the army fighting on the battlefield often depended on them.

- Who watched the battle from the ramparts and knew when to shoot arrows or throw stones? – Someone who didn't "sleep on the job"! Someone who was absolutely reliable, attentive, aware of the enemy's tactics, submissive to the king, loyal to his comrades–in–arms.

Practical lessons

Dear Reader,

I know that you have already clearly understood that we can learn some very valuable lessons from this army of King Uzziah, his 2,600 senior officers, and the remaining 307,500 elite soldiers. Let's reflect on just a few of these lessons and apply them to our individual lives and as members of the Church, soldiers of Christ:

- *Every child of God, whether he realizes it or not, is a soldier in the Lord's holy army. If each of us would strive to be an elite soldier like King Uzziah, we would have more victories, both individually and collectively (as a local church), and suffer fewer defeats;*
- *Each soldier has his own specialty, weapon of choice, and discipline. They must concentrate on what they do best and always try to improve their skills. In comparison, the Holy Spirit has given each child of God one or more gifts (specialization(s)) that require development, training, practice, dedication, effort, and consecration to reach the highest level of service and excellence;*
- *Soldiers watch each other's backs unless they are traitors;*
- *Soldiers remain in battle until the end, unless they become deserters, abandoning their post and their comrades;*
- *Soldiers put their King first, then their brothers in arms, and finally themselves. May the Lord allow His children to do the same: put King Jesus first in their lives, take care of their brothers in arms, and only then pursue their own interests if it honors the Lord;*
- *Weapons must always be aimed at destroying and neutralizing the enemy, never at wounding soldiers of the same army! We must be*

able to recognize who is behind the problems that arise before us, or who is trying to divide and rule, or who is trying to trap God's field, to send the "birds to eat the seed" as soon as it is sown, to sow tares among the wheat, and so on. This is the enemy that must be neutralized and driven out.

- *Just as soldiers long to please their general, we should spare no effort to please the One who saved us and enlisted us in His army! Let's follow Paul's good advice to Timothy, his son in the faith: "Suffer hardship with me, as a good soldier of Christ Jesus. No soldier in active service entangles himself in the affairs of everyday life, so that he may please the one who enlisted him as a soldier." (2 Timothy 2:3–4)*

Words for a tired warrior

Why is your heart downcast and weary?
Why is discouragement overcoming you?
When at the same moment
Is the peace of heaven only a prayer away?

Why don't you believe?
Why don't you see God's way in tribulation?
Why don't you cry out?
Why don't you cry out
For the name that is above every name?

Why did you let your fatigue set in and your faith become numb?
Why didn't you stand up and raise the sword of the Spirit to fight
and win?
Why were your loins not girded with truth?
Why did you let yourself be bound again?
When you had already tasted liberation AND freedom?

Where have you forgotten the breastplate of righteousness?
Where have you put down the helmet of salvation? Why are you
surprised to find yourself defeated, fallen, confused, lost?

Where is the peace of
Walking in the Gospel?
Where is the faith that, like a shield, protects you from the flaming
arrows?

Why do you silence the pain
And keep the fear in the depths of your soul?
Why don't you break out in faith to fight ?
In spite of all that opposes you?
Have you forgotten the vile opponent?
You can't let him win!
Do you think he's resting?
Or is he calling a truce?
He doesn't make an alliance,
He doesn't give way to good,
There is no truth or faithfulness in him!
All in him is darkness without light!
But you know he trembles
And retreats
As you proclaim the name of Jesus in faith!

Don't stop now! Don't leave the path of service to your king!
Rise again, brave soldier! Only before the Lord will you bow down.
Put on your armor,
Take up thy sword, and hold fast thy shield.
The Lord of armies goes before you and is also your rear guard!

He will direct your course and your spirit,
Like a pillar of smoke and a cloud in the desert. He will guard your
steps
Like a pillar of fire in the dark night. When you fall asleep, he's awake.
If you move away, he'll take you in his arms and bring you closer.

Under the scorching sun
He is shade and rest for the marching warrior. In the cold of the

night, in the wilderness, in the lonely place,
In the trenches of war,
In the darkness of any battle,
He is light and he is fire that encourages and warms, that animates
and protects from enemy attack.

Rest a while, weary warrior, but awake again, refreshed!
Put on your armor and prepare for war! Don't let evil conquer the
land that is rightfully yours!

Raise your sword, faithful warrior!
Raise the banner with the king's name on it! And after all, after the
battle, after this life,
You will go to the Father's house where you can finally rest.

But now it's time to start over! You're still here – there's a plan for
you! I want to look to my side
And see you there with me – not weak and tired, but an armed warrior,
Ready to win!

A moment alone with God

Lord, I understand that the moment you saved me, I joined your army.

I acknowledge that I have not always been a good soldier for Christ.
Help me to be disciplined and diligent, to be ready for whatever you
want me to do. Help me to develop the gifts and talents you have
entrusted to me. Help me to keep my armor clean and ready for the
good fight of faith. Renew my strength and send me out again, for
this time I want to follow you.

In the name of the Lord Jesus Christ.

Part III

Men who worship

Chapter 1
Male musicians, singers and worshippers

- Historical data -

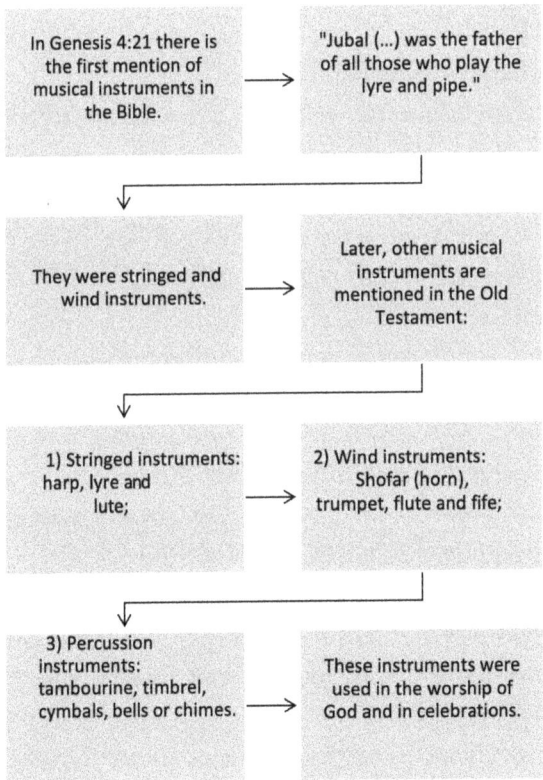

In Genesis 4:21 there is the first mention of musical instruments in the Bible.

"Jubal (...) was the father of all those who play the lyre and pipe."

They were stringed and wind instruments.

Later, other musical instruments are mentioned in the Old Testament:

1) Stringed instruments: harp, lyre and lute;

2) Wind instruments: Shofar (horn), trumpet, flute and fife;

3) Percussion instruments: tambourine, timbrel, cymbals, bells or chimes.

These instruments were used in the worship of God and in celebrations.

Jubal was the son of Lamech, a descendant of Cain. He was the inventor of two musical instruments: the harp and the flute. There are some questions about music:

1. Was there music before musical instruments?
2. Is God the author of music? Or was it Jubal who composed the first melodies heard on earth?
3. What inspired Jubal to invent the first instruments?

I don't know if you agree with me, but I think we can all hear music in nature! I'm not just talking about the pleasant trills of birds, but also, for example, the music of water flowing downstream or crashing over waterfalls, whether small or imposing, the crashing of waves on the beach, the rain drumming out different sounds depending on the hardness or malleability of the surfaces on which it falls! And the melody of the wind as it moves through the trees of a forest! Or just dancing through the leaves on a still day. The rumble of thunder after the metallic sound of lightning.

The sounds of animals, when they communicate with each other or announce their presence, fill our world: the frightening roar of the lion, the chilling "laugh" of the hyena, the roar of the elephant, the bleating of a sheep, even the croaking of frogs or the buzzing of bees over flowers – they are all part of the symphony of the animal kingdom.

Studies carried out on dolphins, the results of which have recently been published, claim that they have different accents depending on the region they come from! What an extraordinary variety! *"He has made everything beautiful in its time."* (Ecclesiastes 3:10, ESV).

We can't forget the melody of a baby's first laugh! It's music to parents' ears: their child is happy and expresses his of her well–being in this way. So we can conclude that singing has been around forever!

In Job 38:7 we read that the stars sang and the angels rejoiced when God created the world. The Lord created music just as He created everything else!

Was Jubal inspired by the sounds of nature to invent the first instruments? Was it a "happy accident" that he discovered that blowing through a reed produced a special sound? Or that a string stretched out would sound different when struck in different places? The Bible doesn't reveal this, but what we do know is very relevant to understanding the role of music in worship.

(Very) Brief note on:

The role of music in the history of the Hebrew people

Music has always been, is and will always be a privileged medium for various purposes. Let's look at some of them:

1. Express and evoke feelings;
2. Communicate ideas;
3. Passing on knowledge and history;
4. Telling personal stories and the history of peoples;
5. Share experiences;
6. Bringing people together;
7. Inspire action;
8. Forming opinions;
9. Teaching;
10. Transform attitudes;
11. Develop hope;
12. Encourage;
13. Remember;
14. Mobilize crowds;
15. Create a revolution;
16. Soothe the soul;
17. To identify a people or nation national anthem;
18. Calm the mood;
19. Set the pace of work;
20. Change the atmosphere;
21. Create a specific atmosphere;
22. Moving;
23. Cheering up;
24. Celebrate a victory;
25. Worship and praise God;
26. Encourage faith;

and much more that the reader will think of that is not mentioned here.

It's common for a song or melody to be associated with a revolution, either because it instigated it or because it recounts the historical events that took place. Every country has its national anthem, a musical composition that evokes the nation's historical achievements and is an official symbol of the state. There is also traditional and reference music that outlines the unique characteristics of a people, identifying and distinguishing them from all others, such as *fado* in Portugal.

Throughout history, a lot of manual labor has been accompanied by music that sets the rhythm, from reapers to fishermen and many other classes of workers.

As a child, I lived with my family in an apartment above a bakery in Mozambique. Bread was kneaded day and night to the sound of a kind of responsive, very lively song. I remember fragments of this song. I loved going to buy some bread and listening to the bakers singing, even though I didn't understand what they were saying.

Thinking now about music as an excellent means of worship, we can say that the Scriptures record many times in the history of the people of the Bible when crowds sang. The first record appears in the book of Exodus, chapter 15:1–18, subtitled *"The Song of Moses"*. The same chapter, in verses 20 and 21, records the *song of Miriam,* Moses' sister, and all the anonymous women who followed her. Miriam led her large choir with a tambourine and her voice. In both cases, they celebrated the triumph of the Lord, His power and His care for Abraham's descendants, fulfilling the promise made to the patriarch.

In the two examples above, it is very likely that this was responsive singing. The former Hebrew slaves were accustomed to singing to punctuate the tasks of each day in Egypt, to praise the Lord, to pass on the traditions of the people and their history to the next generation, to support each other, and to sing about their hope for better days to come.

Egyptian civilization was highly developed in many areas, such as science, literature, and the arts, including music. The Egyptians were polytheists and believed that Thoth, the "god of knowledge and wisdom," had invented music and that Osiris, another god, "lord of the dead," used it to "civilize" the world. Their first instruments were the flute and the harp, but they had already sung music with just their voices and clapping. Music was present at their many celebrations. So it's not surprising, that after more than 400 years of captivity, the Israelites brought with them some influence (or a lot of influence) from the music of Egypt.

When the people of Israel made the golden calf in the desert while Moses was in deep communion with God on Mt. Sinai, there was music and dancing in the camp, similar to the idolatrous feasts in Egypt .

When Moses came down from the mountain with the two tablets of the law, he heard *"the sound of singing"* (Exodus 32:18) and was outraged to see the people worshiping the calf. The influence of Egypt was very evident among the Hebrews:

"So the next day they rose early and offered burnt offerings, and brought peace offerings; and the people sat down to eat and to drink, and rose up to play." (Exodus 32:6)

Throughout Israel's history, music appeared in various celebrations or as a cry or lament, most clearly in the Psalms,§ which were not just songs but also lyric poetry. The Book of Psalms became the hymnal of the Hebrews. One of them, Psalm 90, was written by Moses; the others were written by King David (73 Psalms), King Solomon (72 and 127), Heman (88) and Ethan (89), Asaph and his descendants, and the Korahites (a family from the tribe of Levi).

We should also consider that several Psalms are prophetic: that is, they contain prophecies that point to the Messiah, the Lord Jesus Christ. After His resurrection, on one of the occasions when He shared a meal with His disciples, He said to them:

---

§ According to scholars, the word psalm comes from the Greek and means "to play (a musical instrument)"; in Latin, the term refers to lyrical poetry.

"These are My words which I spoke to you while I was still with you, that all things which are written about Me in the Law of Moses and the Prophets and the Psalms must be fulfilled." (Luke 24:44)

So the book of Psalms is not only a book of music and poetry, but also a book of prophecy.

Singers, composers and players: The example of David

"Now when David reached old age, he made his son Solomon king over Israel. And he gathered together all the leaders of Israel with the priests and the Levites.

The Levites were numbered from thirty years old and upward, and their number by census of men was 38,000. Of these, 24,000 were to oversee the work of the house of the LORD; and 6,000 were officers and judges, and 4,000 were gatekeepers, and 4,000 were praising the LORD with the instruments which David made for giving praise." (1 Chronicles 23:1–5)

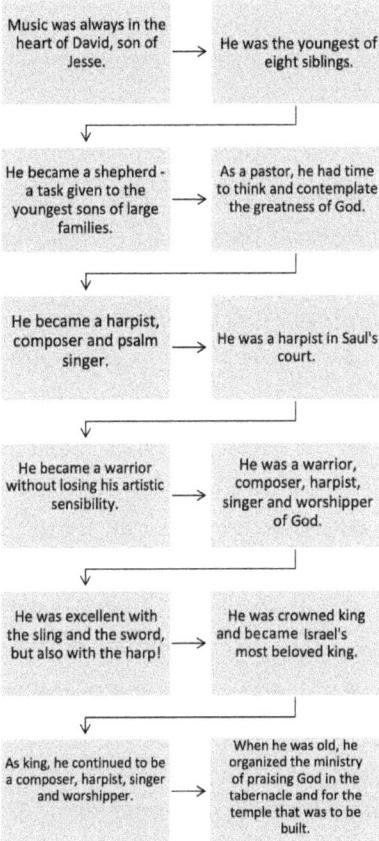

Music was always in the heart of David, son of Jesse. →	He was the youngest of eight siblings.
He became a shepherd - a task given to the youngest sons of large families. →	As a pastor, he had time to think and contemplate the greatness of God.
He became a harpist, composer and psalm singer. →	He was a harpist in Saul's court.
He became a warrior without losing his artistic sensibility. →	He was a warrior, composer, harpist, singer and worshipper of God.
He was excellent with the sling and the sword, but also with the harp! →	He was crowned king and became Israel's most beloved king.
As king, he continued to be a composer, harpist, singer and worshipper. →	When he was old, he organized the ministry of praising God in the tabernacle and for the temple that was to be built.

"Then David spoke to the chiefs of the Levites to appoint their relatives the singers, with instruments of music, harps, lyres, loud–sounding cymbals, to raise sounds of joy." (1 Chronicles 15:16)

King David, himself a composer, singer, and musician, entrusted the kingdom to his son Solomon before entering God's presence. He organized various ministries already established in the tabernacle that would be expanded and developed in the temple, including the ministry of worship. Not only did he write many psalms, but he also composed the music for them to be sung.

In the second book of the prophet Samuel, chapter 22, we find a record of one of David's most beautiful songs, *"in the day that the LORD delivered him from the hand of all his enemies and from the hand of Saul"* (2 Samuel 22:1). He had an extraordinary understanding of God's tremendous majesty, His deliverances, His perfection, and His goodness. For all of this, he concluded his song with these words:

> *"I will give thanks to You, O LORD, among the nations,*
> *And I will sing praises to Your name.*
> *"He is a tower of deliverance to His king,*
> *And shows lovingkindness to His anointed,*
> *To David and his descendants forever." (2 Samuel 22:50–51)*

Regarding the temple itself, David had already gathered a large quantity and variety of materials necessary for its construction, and he also had the plans for the work. However, the Lord did not allow him to build the House; instead, He granted this privilege to King Solomon, his son and successor.

With regard to those who would serve in the temple – their respective tasks and positions – David organized everything wisely. His experience as a warrior, commander of armies, servant of God, worshipper, and ruler was an invaluable asset that he passed on to Solomon. The way he accepted God's will regarding the construction of the temple was exemplary. Solomon himself bore witness to this on the day the temple was inaugurated:

> *"Now it was in the heart of my father David to build a house for the name of the LORD, the God of Israel. But the LORD said to my father David, 'Because it was in your heart to build a house for My name, you did well that it was in your heart. Nevertheless you shall not build the house, but your son who will be born to you, he will build the house for My name.' Now the LORD has fulfilled His word which He spoke; for I have risen in place of my father David and sit on the throne of Israel, as the LORD promised, and have built the house for the name of the LORD, the God of Israel." (1 Kings 8:17–20)*

The way David lived out his last years – ensuring the safety of his people, delivering them from enemies, and providing Solomon with a stable kingdom – reveals the greatness of a man after God's own heart, despite his failings, for which he wept bitterly and repented before the Lord.

In the context of organizing the ministries related to the temple that Solomon would build, David summoned *"the princes of Israel, the priests, and the Levites"* – a small group that remains anonymous to us but was well known to the king, and even more so to God!

Among the Levites he appointed men to oversee the work, officers, judges, gatekeepers, and singers! 4,000 singers! And not just singers, but musicians – men who could praise the Lord with the instruments the king had ordered to be made!

For David, worship and music went hand in hand. Singers and musicians collaborated on the greatest work of all – the worship of God! The house of prayer would be magnificent, but even more remarkable would be the quality of the praise. Music played with a variety of instruments, 4,000 voices, prayers, and prophecies accompanied by harps, lutes, and cymbals would all be integral to the ministry!

""Moreover, David and the commanders of the army set apart for the service some of the sons of Asaph and of Heman and of Jeduthun, who were to prophesy with lyres, harps and cymbals..." (1 Chronicles 25:1)

Asaph, Heman, and Jeduthun are three names recorded among the worship leaders. Many others remain unknown, but all of them – 4,000 singers, along with musicians and leaders – produced such praise and worship on the day of the temple dedication that the glory of the Lord descended and filled the house! Look with me, dear reader, at the biblical account of this event and witness how the Lord responded to the music, the singing, and the worship:

"and all the Levitical singers, Asaph, Heman, Jeduthun, and their sons and kinsmen, clothed in fine linen, with cymbals, harps and lyres, standing east of the altar, and with them one hundred and twenty priests blowing trumpets in unison when the trumpeters and the singers were

to make themselves heard with one voice to praise and to glorify the LORD, and when they lifted up their voice accompanied by trumpets and cymbals and instruments of music, and when they praised the LORD saying, "He indeed is good for His lovingkindness is everlasting," then the house, the house of the LORD, was filled with a cloud, so that the priests could not stand to minister because of the cloud, for the glory of the LORD filled the house of God." (2 Chronicles 5:12–14)

It was some 3,000 years ago that the temple was dedicated and consecrated to the Lord. How we need the visitation of God's glory in our worship services!

Worshipper

When a man chooses to be a worshipper of God, it carries more weight and meaning than we can put into words. Consider this passage, dear reader:

"But an hour is coming, and now is, when the true worshipers will worship the Father in spirit and truth; for such people the Father seeks to be His worshipers. God is spirit, and those who worship Him must worship in spirit and truth." (John 4:23–24)

"Yet a time is coming and has now come when the true worshipers will worship the Father in the Spirit and in truth, for they are the kind of worshipers the Father seeks. God is spirit, and his worshipers must worship in the Spirit and in truth." (John 4:23–24, NIV)

Let's highlight some expressions from this text as presented in two different translations of the Bible:

- *"Yet a time is coming and has now come"* – in other words, it is time! Now is the time. We want to be found worshiping when the Lord comes! Let us make worship our lifestyle. We want to worship and praise the Lord in such a way that His presence fills the place where we are and our hearts overflow with joy!
- *"true worshipers"* – as opposed to false worshippers, those who worship to be seen, those who worship because that's what

everyone is doing at the moment, those who worship only in the church, forgetting the omnipresence of the Lord! As opposed to those who worship with words, but not with their whole heart; as opposed to those who worship when everything goes well for them but are left without music in their soul in the days of great battles, tribulations and pain. If only they knew the power of music! If only they knew the praise of war, the true worship, the praise that is all about our glorious and almighty God! Not *praise–turned–testimony,* the *praise–of–appeal–to–salvation,* the *local–church–traditional–praise,* or the *emotion–exploiting praise.* No, but praise that lifts us into the holy presence of the Lord! Praise that raises us to the level of child–servants and consecrated worshippers – and keeps us in that place, inside and outside the church, every day of the week!

- *"worship in the Spirit and in truth"* – simply by obeying this commandment: *"You shall love the lord your god with all your heart, and with all your soul, and with all your mind, and with all your strength."* (Mark 12:30); without using God's name in vain; without lying! Thinking and rethinking every word, expression, and attitude during praise!

- *"in the Spirit and in truth, for they are the kind of worshipers the Father seeks"* – God the Father seeks those who worship Him with all their heart, all their soul, all their mind, all their strength, all their truth, and all their spirit, guided by the Holy Spirit. He desires people who are consistent in their worship and steadfast in their consecration – not perfect people, but those who recognize that they are sinners redeemed by the infinite and incomprehensible love of the Almighty!

- *"must worship"* – it is necessary to worship; it is necessary to sing, compose, play worship songs that can be accompanied by an invisible choir of angels! Songs inspired by God Himself!

Appeal

Here I make a solemn appeal to all those who know music, who can sing or play musical instruments, recalling the words of the Apostle Paul to Timothy, his son in the faith, a companion in journeys and trials, so that the Gospel may go further and reach more people:

"Do not neglect the spiritual gift within you (...)" (1 Timothy 4:14)

"For this reason I remind you to kindle afresh the gift of God which is in you (...)" (2 Timothy 1:6)

> *I urge you, for the sake of the Lord, to awaken (or reawaken) the gift of God that is within you, and do not neglect it in any way!*
>
> If you have the gift of playing an instrument – play it!" ! If you have the gift of singing – sing! If you don't have a specific gift related to music, leave that place to those who do! Be pragmatic! Think of the good of all and the glory of God first!

Here are some suggestions and practical advice that may be helpful to those who serve the Lord through music or song. If you recognize any of the following in you, please mark them and discuss them with the Lord and the leadership of your local church.

1. *If you are part of a worship group, submit to the leadership!* If you disagree with a decision, have the courtesy to discuss the matter privately with your manager. Don't make snide remarks. Don't belittle anyone. Remember that we are all different and that God has a special purpose for each of us.
2. *Don't sing or play to be noticed, but only to worship the Lord!* The worship group will give you many opportunities to learn to be humble.
3. *Build melodious harmonies within your group and don't be dissonant!* Be generous with the other musicians and singers, be a facilitator, a lever and not a stumbling block.
4. *Attend every rehearsal on time with a willingness to learn and*

serve – to be blessed and to be a blessing! If you can't be there, let your worship leader know it before handand explain the reason for your absence.

5. *Don't hinder the work of others!* If you haven't been to rehearsal, don't sing or play just because you think you "know" the song! The audience will notice that you're not at your best. The rest of the group notices it too, and you even more so.

6. *Be humble and always learn!* Children learn a lot because they are humble and curious; they really want to know! Do as they do! Learning requires: the desire and intention to learn; effort and commitment; attention and dedication; work and repetition.

7. *Be humble enough to recognize that you need help!* Be teachable! No one knows everything, and together we are better and stronger.

   "If anyone supposes that he knows anything, he has not yet known as he ought to know..." (1 Corinthians 8:2)

8. *Acknowledge your mistakes, ask for forgiveness, correct them, and keep learning!* It's a long career; there will be failures. Don't get knocked down or stuck. Get back up, shake off the dust, treat your wounds, set your sights on the finish line and try again!

9. *Pray with the group!* People who pray together become closer, share the same space and breathe the same atmosphere at the throne of grace. They often become friends for life! Problems are solved and tensions disappear when the group bends the knee and each person brings their request before God.

10. *Pray alone before singing, playing, speaking or praying in public!*

11. *The Lord will honor your dependence on Him and your faith. "But you, when you pray, go into your inner room, close your door and pray to your Father who is in secret, and your Father who sees what is done in secret will reward you." (Matthew 6:6)*

Dear reader,

Let's be thankful to the Lord for every opportunity He gives us to serve Him or to be useful to someone else. If the Lord has given you a gift, he has given you a mission, therefore.

I invite you to apply to yourself the words of encouragement that God spoke to Joshua, recorded in chapter 1, verse 9, of the book that bears his name:

"Have I not commanded you? Be strong and courageous! Do not tremble or be dismayed, for the LORD your God is with you wherever you go." (Joshua 1:9)

A moment alone with God

Heavenly Father, if you send me, I will come.
When I'm afraid, I'll call on Your name,
I'll rest in You and trust in You.
When I fail, give me your forgiveness,
Lift me up, take me by the hand
And lead me where you want me to go.
If you send me, I won't go if you don't come! I can't run the race
    alone! But if you follow me along the way, I'll have skill, agility
    and strength, I'll be joyful and swift
As a deer as I climb the hill of worship.
And there, on high ground, I'll have a better view:
I'll see more of the path I've already begun.
I bow before You in holy submission!

Prayer:

Lord, I come to worship You for all that You are and for all that You have done for me. I thank You for welcoming me as Your child and for being my Heavenly Father!

I ask You to forgive me for not being as diligent as I should have been in using the gifts You entrusted to me. I don't want to look back some day and regret the many things I left undone throughout my life, simply because I always found excuses not to do them!

I ask You, Almighty God, to rekindle the flame of Your love within me! Renew Your anointing on my gifts and grant me opportunities to exercise them.

By faith – and with the authority of the name of the Lord Jesus Christ – I now break down all the barriers that have prevented me from obeying You and serving You as You deserve!

By faith, I declare that I am free – now – to honor You with each one of my gifts and each one of my talents. I have nothing that I have not received from You, not even the gift of life. I give You what I am and what I know. I surrender my will to You and pray that You will govern it every day. In the name of the Lord Jesus Christ.

Chapter 2
7000 hidden worshippers

- Historical data -

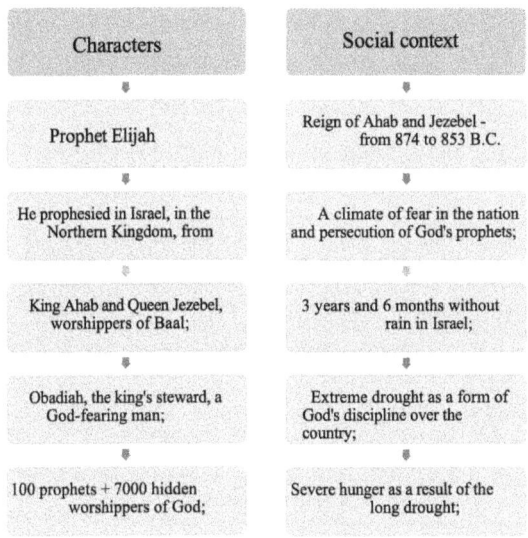

Characters	Social context
⬇	⬇
Prophet Elijah	Reign of Ahab and Jezebel - from 874 to 853 B.C.
⬇	⬇
He prophesied in Israel, in the Northern Kingdom, from	A climate of fear in the nation and persecution of God's prophets;
⬇	⬇
King Ahab and Queen Jezebel, worshippers of Baal;	3 years and 6 months without rain in Israel;
⬇	⬇
Obadiah, the king's steward, a God-fearing man;	Extreme drought as a form of God's discipline over the country;
⬇	⬇
100 prophets + 7000 hidden worshippers of God;	Severe hunger as a result of the long drought;

In this chapter, we will learn about the possibility of worshiping God in secret because of religious persecution. We will remember the Christians who continue to be persecuted for their faith – millions of anonymous people who have been persecuted, imprisoned, discriminated against, and killed for believing in the Lord Jesus Christ in the past centuries, right up to our own day, including the

20th century (considered one of the worst periods of persecution of Christians in the world) and in the 21st century.

Below is a brief outline of the characters who are part of the historical story being studied, as well as the context in which the events took place.

Biblical narrative

(in the New Transformation Version)

> "After a long time, in the third year, the word of the Lord came to Elijah: "Go and present yourself to Ahab, and I will send rain on the land." So Elijah went to present himself to Ahab.
>
> Now the famine was severe in Samaria, and Ahab had summoned Obadiah, his palace administrator. (Obadiah was a devout believer in the Lord. While Jezebel was killing off the Lord's prophets, Obadiah had taken a hundred prophets and hidden them in two caves, fifty in each, and had supplied them with food and water.) Ahab had said to Obadiah, "Go through the land to all the springs and valleys. Maybe we can find some grass to keep the horses and mules alive so we will not have to kill any of our animals." So they divided the land they were to cover, Ahab going in one direction and Obadiah in another.
>
> As Obadiah was walking along, Elijah met him. Obadiah recognized him, bowed down to the ground, and said, "Is it really you, my lord Elijah?"
>
> "Yes," he replied. "Go tell your master, 'Elijah is here.'"
>
> "What have I done wrong," asked Obadiah, "that you are handing your servant over to Ahab to be put to death? As surely as the Lord your God lives, there is not a nation or kingdom where my master has not sent someone to look for you. And whenever a nation or kingdom claimed you were not there, he made them swear they could not find you. But now you tell me to go to my master and say, 'Elijah is here.'

I don't know where the Spirit of the Lord may carry you when I leave you. If I go and tell Ahab and he doesn't find you, he will kill me. Yet I your servant have worshiped the Lord since my youth. Haven't you heard, my lord, what I did while Jezebel was killing the prophets of the Lord? I hid a hundred of the Lord's prophets in two caves, fifty in each, and supplied them with food and water. And now you tell me to go to my master and say, 'Elijah is here.' He will kill me!"

Elijah said, "As the Lord Almighty lives, whom I serve, I will surely present myself to Ahab today."

So Obadiah went to meet Ahab and told him, and Ahab went to meet Elijah. When he saw Elijah, he said to him, "Is that you, you troubler of Israel?" (1 Kings 18:1–17).

The Lord's words to the prophet Elijah some time later:
"Yet I will leave 7,000 in Israel, all the knees that have not bowed to Baal and every mouth that has not kissed him." (1 Kings 19:18, NVT)

Persecution in the days of the prophet Elijah

The ascent of King Ahab to the throne, married to Jezebel, the daughter of Ethbaal, king of the Sidonians, led to the establishment of idolatry in Israel, particularly the worship of Baal. To consolidate this new religious state, intense persecution was carried out against the prophets of God – the God of Abraham, Isaac, and Jacob – the Creator and only Lord of all that exists.

When we read the commandments that God gave to Moses, we find these solemn words from the Lord to His people:

"God spoke all these words:

I am God, your God,
who brought you out of the land of Egypt,
out of a life of slavery.

No other gods, only me.

No carved gods of any size, shape, or form of anything whatever, whether of things that fly or walk or swim. Don't bow down to them and don't serve them because I am God, your God, and I'm a most jealous God, punishing the children for any sins their parents pass on to them to the third, and yes, even to the fourth generation of those who hate me. But I'm unswervingly loyal to the thousands who love me and keep my commandments." (Exodus 20:1–6, MSG)

The Lord is the only God and does not allow idol worship – it's as simple as that! There is no refuting the clarity of these commandments. I chose to quote this portion of Scripture from The Message because it leaves no room for doubt. The consequences of worshiping entities other than the Lord are also made explicit.

The persecution of the worshippers of God the Almighty was like a sieve sifting the faith of the Israelites:

- A very large number chose to fear the king and queen, to bow down and worship Baal;
- others were killed for obeying the Lord and refusing to worship idols;
- some fled to remote areas;
- others went into hiding;
- still others hid their faith and tried to survive by continuing to worship the Lord in secret.

This state of affairs had a very strong impact on the nation, which became divided and walked *"limping between two different opinions".* *(1 Kings 18:21, ESV)*

The discipline of the Lord then fell upon Israel. Elijah was sent to King Ahab with a stern message from God:

"Now Elijah the Tishbite, who was of the settlers of Gilead, said to Ahab, 'As the LORD, the God of Israel lives, before whom I stand, surely there shall be neither dew nor rain these years, except by my word.'" (1 Kings 17:1)

By appearing before the king, Elijah knew he was risking his life. But he couldn't fail to deliver the Lord's message: days of drought and famine were coming as God's way of disciplining the nation, which had fallen into idolatry under the leadership of Ahab and Jezebel. God's word did not tell the king how many days, months, or years the drought would last. There was no time to make preparations, to divert water from the rivers to fill the cisterns, to store food, to gather pasture for the cattle – nothing! The drought began then and there. Suddenly. And the king wasn't prepared...

When Jezebel heard the news, she ordered that Elijah be killed! But her long, evil arm would never reach him. The Lord, under whose wings the prophet hid, sent him to a safe place, a mixture of refuge and retreat, a quiet place where the sounds of water and birds were his company. He had time to meditate, to learn from God and to speak with Him.

"Go away from here and turn eastward, and hide yourself by the brook Cherith, which is east of the Jordan. It shall be that you will drink of the brook, and I have commanded the ravens to provide for you there." So he went and did according to the word of the LORD (...) (1 Kings 17:3–5)

If we reflect a little on Elijah's behavior, reader, we can conclude that:

- Hiding was an act of prudence. It was not (yet) time to confront the forces of evil that were at work in Israel, under the protection of the king and queen. Elijah's mission was far from being complete, so he had to live to see it through – to bring the nation back to God!
- Obeying the Lord was an act of faith. Only a fool would believe that some crows would come to bring bread and meat to a lonely man on a deserted creek! Or a man of extraordinary and unshakable faith. Such was the case!
- Staying at the Kerite torrent (or stream) was an act of wisdom. Water would be scarce throughout the area. There was water at Kerite. He needed to stay there until God told him to leave.

A new message for the prophet

The time had come for Elijah to leave his safe place where the sky was his roof, the vegetation his walls, and the Word of God his weapon. The riverbed was dry. The ravens would not return to bring him food, they would not return to drink from the same water that the prophet drank from. He shared the water, they shared the bread and the meat. What a rare symbiosis! But it was over. It was time to move on.

Reader, it's important to recognize when the work we are doing for God has come to an end and it's time to move on. By doing so, we will avoid the painful maintenance of dying ministries that bear no fruit and do not honor the Lord. Stay vigilant!

This time God sent Elijah to Zarephath. The city had once been large and important on the Mediterranean coast of Lebanon. It belonged to the Sidonians, where Jezebel's father reigned, the queen who wanted to destroy all of God's prophets! Although it was outside the borders of Israel, drought and famine also reached this region.

In Zarephath, Elijah would learn some very valuable lessons! Let's look at three of them:

1. The great lesson of humility – he would be supported by a poor, widowed woman, which was contrary to the basic principles of society at the time. Being supported by a woman was already a delicate matter, and if she was a widow, the situation became even more challenging. Adding the fact that she was extremely poor made the circumstances overwhelming! Elijah was a strong man who had to learn to make himself weak so that God's power could be manifested through him. It was a significant test of character, which the prophet overcame with remarkable success.

2. The great lesson of expecting great things from the Lord and continuing to trust Him, regardless of the circumstances – rather than merely receiving help from a poor widow, he became the solution to the extreme hunger faced by that

woman and her son. He became a blessing to that small family because God, for his sake, provided the portion of flour needed for each day in the widow's pot and enough oil for the bread in her jug! In this way, he became the blesser, while the widow and her son were the blessed ones – a new symbiosis.

Just as Elijah had been able to pray at Kerith's dwindling stream: "Give me the food I need to live today," so now, facing an empty pot and a jug of oil without a drop in it, he would continue to depend on the Almighty for his daily portion! The Lord knew he was there – He had sent him there to provide what was needed.

Notice, reader, how much easier it would have been if God had given him several bags of flour and several jugs of oil at once! But He didn't. He gave him, day by day, enough to satisfy his hunger and the hunger of the widow and her son. How the Lord blesses us when we are a blessing to others!

3. The great lesson of knowing the unlimited power of the Lord – when Elijah, by His infinite power, was the instrument God used to bring the widow's son back to life!

In that poor house, sheltered under an old roof and hidden from Ahab and Jezebel, unnoticed by Ethbaal, Jezebel's father, Elijah remained in complete safety and anonymity until the time came to take another step forward and continue his mission.

A new message for the king - Part I

Now it happened after many days that the word of the LORD came to Elijah in the third year, saying, "Go, show yourself to Ahab, and I will send rain on the face of the earth." (1 Kings 18:1)

The message was from God – the only God – and it was a message of hope. Would the king give him time to speak or would he have the prophet executed where they met?

Elijah was not afraid. He left Zarephath and made his way back to Samaria. He had a word from the Lord to deliver to the king. He would deliver it with all his heart!

Meanwhile, King Ahab was looking for pasture and water for some of his horses and mules. He summoned his butler, Obadiah, the steward of the palace, to accompany him on this search. Two groups were organized: one to follow the king and the other to follow Obadiah. They set off in different directions, searching intently for water and pasture, with very little success.

At one point along the way, Obadiah saw a man walking toward him, alone, with a firm and determined step... Could it be the prophet Elijah? Where could he have been? How could the king not have found him earlier? What was he doing there?

The conversation between the two men was brief but enlightening. Elijah had a message for the king. Obadiah was to go and tell him. Elijah demanded an emergency meeting with Ahab! He had news from God.

100 hidden prophets = 100 anonymous worshippers

Obadiah, in turn, told the prophet Elijah about the condition of the nation:

- Ahab had searched everywhere for Elijah, including Israel's neighboring kingdoms;
- Ahab intended to kill Elijah;
- Jezebel had put to death all the prophets of the Lord she had found.

And then came the best part: Obadiah had managed to save 100 prophets from the hands of Ahab and Jezebel!

"Has it not been told to my master what I did when Jezebel killed the prophets of the LORD, that I hid a hundred prophets of the LORD by fifties in a cave, and provided them with bread and water?" (1 Kings 18:13)

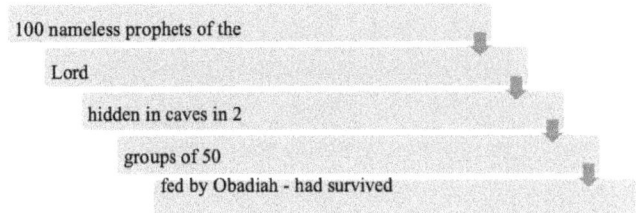

100 nameless prophets of the
Lord
hidden in caves in 2
groups of 50
fed by Obadiah - had survived

100 prophets whose names the Lord does not reveal to us in Scripture were as faithful to the Lord as those who lost their lives, as faithful as those who did not bow down to idols, as faithful as those who escaped the edge of the sword (Hebrews 11:14), and as faithful as Elijah!

By faith, 100 men lived hidden in caves under inhuman conditions, depending on Obadiah – a godly man in a social position that allowed him to save and preserve the lives of these prophets. How many more of God's people would Obadiah have liked to save, but he couldn't! For 100 men, however, the king's was the difference between life and death.

Reader, please notice steward the following:

The name Obadiah means "one who serves God" and "worshipper of the Lord. "Other acceptable meanings of this name are "obedient" and "dedicated". In the brief account of Obadiah's efforts to save the 100 prophets, his attitude toward Elijah, and his ability to endure continuing to serve King Ahab as an "undercover agent" of God, we see his exceptional character, wisdom, resourcefulness, humility, and courage that earned him the Lord's trust for the work he was allowed to do. May God grant that we are all worthy of the same trust!

A new message for the king - Part II

Elijah did come to meet King Ahab. After a brief and bitter introductory dialogue, the prophet challenged the king to a kind of spiritual duel between Baal and the Lord on Mount Carmel.⁵ More than a duel between a powerless idol and the Almighty God, it would be a battle between the forces of darkness and the army of the Lord!

On the appointed day, at the appointed time, the darkness was represented by 450 prophets of Baal and 400 prophets of the idol Asherah. The Lord was represented by one man: Elijah. The people of Israel came as spectators.

"Elijah came near to all the people and said, "How long will you hesitate between two opinions? If the LORD is God, follow Him; but if Baal, follow him." But the people did not answer him a word." (1 Kings 18:21)

Out of fear or disbelief, the people did not respond to Elijah's call. Essentially, everyone was on the mountain to see who God really was.

"I alone am left of the prophets of the Lord, (...)" – said Elijah, not mentioning the 100 men that Obadiah had hidden and supported. But not even the fact that he was alone drew the sympathy of the people.

Two altars were then prepared for the sacrifice of a bullock – one in honor of Baal, the other in honor of the God of Israel. There were some ruins left of what had once been an altar to the Lord on Mount Carmel. Elijah rebuilt the old altar with 12 stones, each representing a tribe of Jacob. He dug a trench around the altar and had it filled with water three times. The bullock and the wood on the altar were also watered three times, and then Elijah prayed. After everything is done, after all the obedience, it is necessary to pray!

Elijah's prayer and the people's reaction:

At the time of the offering of the evening sacrifice, Elijah the prophet came near and said, "O LORD, the God of Abraham, Isaac and Israel, today let it be known that You are God in Israel and that I am Your servant and I have done all these things at Your word. Answer me, O LORD, answer me, that this people may know that You, O LORD, are God, and that You have turned their heart back again." Then the fire of the LORD fell and consumed the burnt offering and the wood

---

⁵ Mount Carmel is located on the coast of Israel, overlooking the sea.

and the stones and the dust, and licked up the water that was in the trench. When all the people saw it, they fell on their faces; and they said, "The LORD, He is God; the LORD, He is God." (1 Kings 18:36–39)

This is how the Lord manifested His presence and power, calling His people back to faith in the one God and Lord!

The last sentence of the message to King Ahab was still missing: Rain will fall on the people and water the dry land!

Now Elijah said to Ahab, "Go up, eat and drink; for there is the sound of the roar of a heavy shower." So Ahab went up to eat and drink. But Elijah went up to the top of Carmel; and he crouched down on the earth and put his face between his knees. He said to his servant, "Go up now, look toward the sea." So he went up and looked and said, "There is nothing." And he said, "Go back" seven times. It came about at the seventh time, that he said, "Behold, a cloud as small as a man's hand is coming up from the sea." And he said, "Go up, say to Ahab, 'Prepare your chariot and go down, so that the heavy shower does not stop you.'" In a little while the sky grew black with clouds and wind, and there was a heavy shower. And Ahab rode and went to Jezreel. Then the hand of the LORD was on Elijah, and he girded up his loins and outran Ahab to Jezreel. (1 Kings 18:41–46)

Elijah said it would rain long before there was a single cloud in the sky. And when the first cloud appeared, he sent word to the king to hurry home lest the rain catch him on the way! Just as he said, to it happened.

In Deuteronomy 18:22, there is a clear recommendation as to how we can tell if a prophet is inspired by the Lord or not:

"When a prophet speaks in the name of the LORD, if the thing does not come about or come true, that is the thing which the LORD has not spoken. The prophet has spoken it presumptuously; you shall not be afraid of him."

Today, as in the past, false prophets appear in many places and in many religious circles. Some have many followers, fill large auditoriums, gain fame, and enthusiastically make bold statements that come from their own will or understanding and end up not being fulfilled. Now, if there is no fulfillment of the prophecy made, it is clear that the person who spoke did not do so at the command of the Lord. Their own words will testify to the falsity or truthfulness of what was said.

7000 worshippers of the Lord

As Elijah's ministry neared its end, the shadows of Jezebel's threats loomed large over his path. They lurked – taking refuge, hiding, and settling in the mind of the great and fearless prophet – transforming him into a frightened man in search of any cocoon where he could curl up and hide. Then he ran away. He had fallen into a deep pit of exhaustion and depression. He felt alone. He was alone. He didn't know the whereabouts of the hundred prophets that Obadiah had spoken of. He wanted to die, but not at Jezebel's hand! He wished to enter eternity by the hand of the Lord, his God. So he fled and asked the Lord to take him. Instead, the Lord fed him, strengthened him, and led him to another mountain – Horeb – where, centuries earlier, He had given the Law to Moses and made a covenant with Israel.

Once on the mountain, Elijah hid in a cave, cowering and frightened like a wounded animal. There the Lord found him, and after having caused the mountain to shake, a very strong wind to blow, the rocks to split, and a fire to break out, He spoke with a "still, small voice," confronting Elijah about his escape without frightening him, with the mercy and kindness that the prophet needed to receive at that moment in his life.

Elijah would have to turn back! He would have to stop feeling sorry for himself and respond! He would have to walk the same path for 40 days until he reached the starting line of the new and final stage of his mission:

- To anoint a new king for Syria;
- To anoint a new king for Israel;
- And anoint a new prophet to take his place.

Moreover, the Lord told him that he was not alone: there were

7,000 men in Israel who had not corrupted their faith or worshiped the idols of Ahab and Jezebel. 7,000 men who bowed the knee only to God and to nothing and no one else! 7,000 men whom God knew perfectly well and who had remained faithful to the Lord through Ahab's reign of terror, injustice and idolatry and through the years of drought and famine! Elijah made his way back, returned to the mission, and carried out every task with precision. He trained his successor, Elisha, taught him the Word, shared his experiences, told him about his encounter with God at Horeb, told him about the still small voice of the Lord... and waited – not for death, but for the rapture that the Lord had revealed to several prophets of his time. And so it came to pass: Elijah was caught up into heaven in a whirlwind, with chariots of fire and horses of fire, without having to cross the valley of the shadow of death! He had accomplished his mission and arrived home.

A brief note on:

The persecution of Christians today

More than 340 million Christians around the world face some form of opposition as a result of their identification with Jesus Christ. (Open Doors Mission)

Let's look together at the list of the 20 countries where Christians are most persecuted today (this data is from January 2021):

1. North Korea
2. Afghanistan
3. Somalia
4. Libya
5. Pakistan
6. Eritrea
7. Yemen
8. Iran
9. Nigeria
10. India
11. Iraq
12. Syria

13. Sudan
14. Saudi Arabia
15. Maldives
16. Egypt
17. China
18. Myanmar
19. Vietnam
20. Mauritania.

The list goes on – it's very long, more than 70 countries – and can be accessed on the websites of Open Doors, Voice of the Martyrs, The Esther Project, and others. These Christian organizations also publish a map of the continents with the countries where persecution is most severe, with the degree of persecution classified as very high, medium or worrying.

If you do a little research, you will find that the Middle East, Central and South Asia, Asia–Pacific, Africa, and Latin America have terribly extensive areas of severe persecution. The reasons for this discrimination and "hunting" of Christians are not the same in all countries: they range from political and dictatorial persecution to religious nationalism, clan oppression, Islamic oppression, corruption, organized crime, and others that we don't need to mention here. It is clear that behind the atmosphere of persecution that contaminates society and oppresses individuals, there is a sinister plan to silence the Church of the Lord Jesus!

While Western Christians sleep soundly, there are millions of brothers and sisters scattered all over the world suffering atrocities that don't even cross our minds. While we complain about the rain or the sun, the cold or the heat, the salaries and the inflation, the lack of attention of the economy or the government to the health of the people, there are millions of families without a decent place to live or to die, without food or water, without clothes or transportation, without medicine or any medical help, without freedom or security.

In general, we don't like to think about these things. We avoid mentioning them. We choose to ignore and not be disturbed by the devastating news of men and women giving their lives for the love of Christ!

There are countries where many children are separated from their parents simply because they are considered incapable of raising them "for the good of the nation" because they are Christians! These children are taken to state institutions (orphanages and asylums) and "raised" to be useful citizens for the state. Bibles are confiscated. Churches are burned. People are tortured, especially pastors and workers whose families, if they survive, are ostracized forever. And we – what are we doing about all this? The West sleeps because real persecution has not yet knocked at its door, though it is on its way. The kind of discrimination that true Christians face in this part of the world is not (yet) comparable to what is happening in other parts of the world.

In Hebrews 13:3 we read these words:

"Remember the prisoners, as though in prison with them, and those who are ill–treated, since you yourselves also are in the body."

"(...) Remember my imprisonment. (...)" – wrote the apostle Paul in his epistle to the Colossians (4:18).

The persecution of Christians has been going on since the first century of the Christian era. We are aware of the crimes that have been committed and the wars that have been fought, supposedly "in the name of Christ" throughout history! But any casual analysis will lead us to the conclusion that these were wars waged by men interested in conquering territories, expanding kingdoms and obtaining dividends, not wars with God's approval! Men acted on their own and sowed much pain and misery.

We are no less aware, however, of the diabolical manipulative forces that have always orbited, and still orbit, the governments of nations, extremist and oppressive religious factions, their sophistries and philosophies wrapped in the cloak of science, intelligence, research, development, tolerance, modernity, progress, and the like! It is estimated that 1 million Christians were killed for their faith between 2001 and 2010, and about 900,000 between 2011 and

2020. Yes, already in the 21st century. Although these numbers are not exact, they are very close to reality. It is more likely that the total number of martyrs will be higher rather than lower than the figures given. It is also possible that those who are persecuted for their faith outnumber those who are counted and publicized. In any case, these figures are unavoidable and serve to give us an idea of the dimensions of the persecution that is taking place against those who believe in Christ and refuse to deny their faith.

The 10/40 Window – a term coined in 1990 by Luis Bush, a missionary, pastor, author, and leader of various Christian movements, to refer to the countries between the 10th and 40th parallels north of the equator – encompasses the majority of nations with no (or very limited) access to the Bible and other Christian resources, with very small Christian communities, where poverty and low quality of life are widespread, and where persecution of the church is constant.

Today, several decades after the term *10/40 Window* was coined, countries in other regions of the world have also become intolerant of the Christian faith, indicating that new fronts in the war against Christianity continue to emerge. As a result, the 10/40 Window no longer encompasses all the areas where our brothers and sisters are losing their jobs, families, homes, freedom, or lives because of their faith. Now is not the time to ignore these facts. On the contrary, it is time to get down on our knees and intercede for the Christians in the 10/40 Window and for all others whom Satan has marked with a "target for slaughter" sign on their backs!

Reader, let's not think that we are safe from all these tribulations, because the Lord Jesus Christ spoke very openly about them, as recorded in the Gospels. While we have the blessing of living in countries that allow us to worship the Lord in freedom, let us take advantage of every moment of meeting with the Church and grow in faith, let us grow a lot, because we don't know what tomorrow will bring!

A moment alone with God

Prayer:

Lord, I ask You to forgive me for being so oblivious to my brothers and sisters in the persecuted church in so many countries!

I ask you to help me intercede for these things:

- for them, so that their faith will not fail;
- for the governments of their nations, that they may change their minds, for You, Lord, can "incline the heart of the king" like one who turns a stream to a new bed;
- for the families who lose their fathers and are left destitute;
- for the children who are taken from their families and placed in institutions, leaving them at the mercy of the executioners of their parents;
- for the needs of these persecuted brothers and sisters that the Western Church has neglected.

Come, Lord, show me in what other ways I can help them. I pray in the name of the Lord Jesus Christ.

Part IV

Ordinary men

Chapter 1
A man in the field

- Historical data -

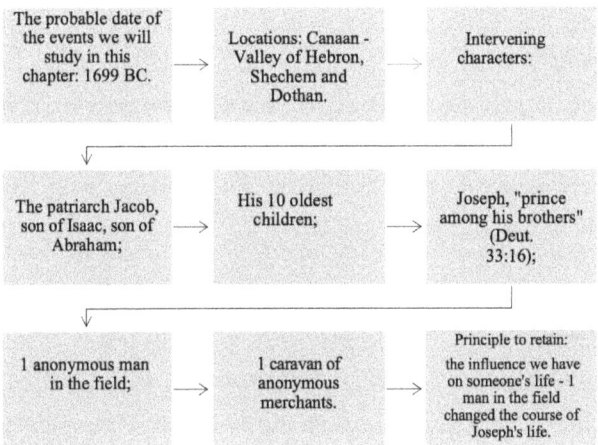

The probable date of the events we will study in this chapter: 1699 BC. → Locations: Canaan - Valley of Hebron, Shechem and Dothan. → Intervening characters:

The patriarch Jacob, son of Isaac, son of Abraham; → His 10 oldest children; → Joseph, "prince among his brothers" (Deut. 33:16);

1 anonymous man in the field; → 1 caravan of anonymous merchants. → Principle to retain: the influence we have on someone's life - 1 man in the field changed the course of Joseph's life.

In the field

I invite you to see the world as a field. Every day we meet many people walking anonymously in the field, just like us. Some seem to know the terrain they're walking on well, they seem to know where they're going, whether the path they're taking is good or bad; they make their own decisions, they follow their own course; others wander the countryside, disoriented, looking for someone or something. They are not satisfied. They search, but they don't find it. They have gone out into the field with a mission, but they don't reach their goal.

This was Joseph, the most beloved son of the patriarch Jacob, son of Isaac, son of Abraham, heir and founder of the nation of Israel.

An ordinary man, an anonymous man, met Joseph and gave him the information he was looking for: the probable location of his 10 brothers with their father's flocks. A simple piece of information. Banal. Anyone could have done the same...

This man had no way of knowing what would happen to Joseph. He had no way of knowing what plans God, the Almighty, had for this 17–year–old boy. He had just entered the story of his life, without realizing the contribution he was making to the direction the nation of Israel would take.

Too often we have the opportunity to help someone, to show them the way, without realizing that we could make a tremendous difference in their life, for their good or for their bad... The next time you meet someone who is lost in life, make sure you contribute to their eternal good: show them the way – tell them about Jesus! Invest some of your time in this mission. One day you'll see that it wasn't a waste of time!

Ready to go

The following biblical text is from Genesis, the first book of the Bible, the book of beginnings, in the New Transformation Version. The choice of this version here in particular is due to the use of expressions that make the text clearer and richer for us readers. When we study and compare different translations and versions of Scripture, we find a greater richness in the text and a broader sense of what is being conveyed to us.

In this excerpt, which we will look at next, several characters are mentioned:

- Jacob, father of a large family of 12 boys and one girl;
- Joseph, the son of his heart, the exemplary boy, was 17 years old at the time of the events we are about to recount;

- Jacob's other sons (we know from the Bible account that there were 10 men who took the flocks to graze away from home);
- An anonymous man who met Joseph in the field while he was Looking for his brothers who were tending their father's flocks;
- God, the invisible orchestrator of Joseph's story.

Israel said to Joseph, "Are not your brothers pasturing the flock in Shechem? Come, and I will send you to them." And he said to him, "I will go." Then he said to him, "Go now and see about the welfare of your brothers and the welfare of the flock, and bring word back to me." So he sent him from the valley of Hebron, and he came to Shechem.

A man found him, and behold, he was wandering in the field; and the man asked him, "What are you looking for?" He said, "I am looking for my brothers; please tell me where they are pasturing the flock." Then the man said, "They have moved from here; for I heard them say, 'Let us go to Dothan.'" So Joseph went after his brothers and found them at Dothan.

When they saw him from a distance and before he came close to them, they plotted against him to put him to death. (Genesis 37:13–18)

Word of mouth

It was reported that a wealthy man from the land of Canaan had lost a son. What a tragedy! He was a good–looking boy, they said. He was only 17. Still a teenager. It was said that he was his father's favorite son, very helpful, intelligent, hardworking, honest, loyal, a real prince...!

His Father is heartbroken! Poor Father! If only he knew for sure what had happened... The other son brought him his blood–stained tunic... They told him that he had been attacked by some kind of beast... This was strange... Possible, of course, but no one found the young man's remains... No one saw any beast prowling around the sheep in the area where the 10 brothers had taken their flocks to graze. However, we must admit that it was not uncommon for predators to suddenly appear... If only they had found Joseph's body, the father and the rest of the family could have buried him, mourned him, and tried to move on with their lives... But that didn't happen... Joseph was a cold case, an unsolved mystery.

There were no witnesses to any of the facts. If anyone knew anything for sure, they kept the secret in the deep vault of their conscience, in the vault of their heart, impenetrable and inaccessible to human eyes. But God knew...

God's eyes followed Joseph the all the way. The encounter with the man in the field was planned, it wasn't a coincidence. The Lord had a purpose in all of this... But what was it?

The Father

This father is now called Israel. In the past, he was Jacob, a supplanter. But in a personal encounter with the angel of the Lord, his name was changed to Israel (Genesis 32:28), the man who *"struggled with God"* or *"the man who sees God."* He was never the same again after that day! He wrestled with God and lived! The Lord wrestled with him within the limits of human weakness, not as the Almighty, but as a man who understands the affliction of another man. *That's how our God is: He takes pity on our weaknesses, remembers that we are dust and does not treat us as our faults deserve, but according to His mercy and His unfathomable and inexplicable love!*

Now, older and weaker, bowed down by the pain of losing Joseph, this man is a father crying out at the feet of another Father, God the Creator, God of all mercy and grace, who would one day see His beloved Son give His life as a ransom for many! God understood the pain of Israel, the struggling man. God didn't tell him that he had an immediate answer to his pleas, but he listened to him passively, like a father listening to another father's lament, like a friend listening to his friend's distress, listening to his tears and sighs, without pointing anything out, without providing quick remedies, the kind that don't really heal. God listened to him and knew that one day all would be well...

The man in the field

It seems impossible to me that news of Joseph's disappearance would not have reached the man who had told him where to find his brothers and flocks! Being from that area, he would have heard the facts about the missing young man. Reader, try to put yourself in this man's shoes and imagine what he might have thought:

"I was the last one to see the young man alive! I sent him to his death without knowing it! I only answered the question the asked me. I told him in which direction his brothers had gone. I couldn't have imagined that, further on, he would be attacked by a beast... I didn't do anything! I didn't touch him! I couldn't hurt anyone! Besides, he looked very distinguished, he was very polite! I can hardly believe what happened to him..."

Have you ever felt like that? Guilty without being guilty?

Yes, guilt can come into the picture. It can come when it's too late to turn back, when the opportunity is long gone, beyond our reach, beyond our horizon. That's when pointed fingers are raised in our direction, as if they were the fingers of our own hands, but they may not be! Look:

- It could be our conscience talking to us, of course;
- It could be the Holy Spirit, if we are children of God, with his *still, small voice,* drawing our attention to some specific aspect of our behavior, our language, choices, thoughts, feelings – a still, small voice, firm and compassionate at the same time, specific and precise like a compass that always points north. It is a voice that always leads to a solution of the problem, to repentance or to a change of life, with the help of the arm of the Lord;
- But it could also be Satan, or one of his agents, accusing you. – They lie to us, crush us, try to convince us that it's all our fault, that we can't do anything right – ever! We're a total failure,

worthless, useless, wretched! This is a voice like that of a *"roaring lion, seeking someone to devour"* (I Peter 5:8), the voice of the accuser, a merciless and destructive voice that does not lead us to repentance or to a change of life. On the contrary, it tries to sink us into our own failure and despair, hammering away incessantly, sucking away our strength until we can no longer resist! And when it catches us in this position of vulnerability, it strikes another blow to bring us down for good! That's how the Evil One works!

Reader, if you're struggling with a feeling of guilt, take the matter to God in prayer. And if you still have a chance to make things right, don't put it off! Do what you can while you still have time and opportunity. Don't carry this burden for the rest of your life. It doesn't have to be that way!

That man may have found himself thinking:

"If I had taken this young man to his brothers, this would not have happened! He would have gotten there safely! It's my fault! With my experience of walking these paths, I could have helped him! I should have realized that he couldn't defend himself against a beast on his own!... I should have thought things through! I should have done things differently!... I should have left my work and gone with him... He might still be alive... It's my fault... I failed a young man who came to me for help... I'm a failure!"

Don't let Satan do this to you! Stand firm in the position and freedom that Christ won for you on the cross of Calvary! Don't be deceived! See, this is what the Word of God guarantees:

"Submit therefore to God. Resist the devil and he will flee from you." (James 4:7)

"Therefore humble yourselves under the mighty hand of God, that He may exalt you at the proper time, casting all your anxiety on Him, because He cares for you. Be of sober spirit, be on the alert. Your adversary, the devil, prowls around like a roaring lion, seeking someone to devour. But resist him, firm in your faith, knowing that the same experiences of suffering are being accomplished by your brethren who are in the world. After you have suffered for a little while, the God of

all grace, who called you to His eternal glory in Christ, will Himself perfect, confirm, strengthen and establish you. To Him be dominion forever and ever. Amen." (1 Peter 5:6–11)

Don't allow yourself to be controlled, except by the God of love! Resist your greatest enemy! Believe that you can overcome through the Lord Jesus Christ because of His name and His perfect work on the cross! Believe and be strong and courageous! The world needs such men!

You know, reader, there are people who pass through your "field" just once. They come across you in a certain place, under certain circumstances, and suddenly there is someone whose life could be affected by a phrase, a gesture, or an attitude on your part! What a unique opportunity! If it's someone who is disoriented, someone who was expecting to meet his "brothers" but ended up in a deserted place and was left alone, pay some attention to him, see what you can do for him. The most appropriate help doesn't always come from the brothers. Sometimes it comes from an anonymous person, at a brief crossroad in life.

We now know what no one knew then

The Bible tells us what happened to Joseph in Genesis 37:19–28. Joseph walked the extra mile to obey his father and find his brothers. In addition to the distance he had already walked, he also walked about 20 miles (or maybe a little more), the distance between Shechem (the place where his brothers were supposed to be) and Dothan (the place where they actually were).

The brothers saw him and quickly devised a plan to get rid of him, of his presumption that he had special dreams, that he was their father's favorite, that he was the "right one" in the family! The plan was simple but terrifying: they would kill Joseph!

But they were not alone in the field. They didn't know it, but the fact was that God was overseeing every step of this young man who had been chosen for a great mission! He wouldn't let them kill him. This caused a division among the ten men: the majority were in favor of Joseph's summary execution, but there was one, Reuben, the eldest brother, who openly opposed the plan of the others. Reuben, a weak link... The chain could break there...

Reuben felt responsible for all his brothers: he was the oldest. So he hatched a plan of his own: find a way for Joseph to escape! he suggested that he could ne thrown into an empty cistern, hoping to buy his persecuted brother some time, if only for a few hours, and later to find an opportune moment to let him escape and return home while trying to divert the attention of the rest of his brothers elsewhere.

It was a well–intentioned plan, but very difficult to execute!

On God's horizon

In Dothan, on the horizon, was the silhouette of a large hut of Ishmaelite and Midianite traders. The second part of God's plan for Joseph's life was unfolding unexpectedly, and it would change the course of his story.

Caravans were generally made up of a mixture of people from different backgrounds who joined together to make long commercial journeys and to protect each other and their goods from both robbery and attack by animals, as well as to support each other and defend themselves against the elements: wind, heat, cold, storms, lack of water and difficulty in finding safe places to stay overnight, fatigue... Most of the members of a caravan were merchants. Whenever possible, loads were transported by boat, across rivers or by sea. When that wasn't possible, they used caravans. According to archaeologists, it was not uncommon for there to be up to 3,000 donkeys in a single caravan, which also meant that there was a very large group of people trading a wide variety of goods.

Nine of the brothers saw the caravan coming. Reuben saw nothing! He could have been tending the flock on his own, or rethinking his plan to save Joseph from death, or perhaps gaining the courage to intervene again to confront the other brothers... We don't know. His position was difficult! All we know is that he wasn't with the others.

This time it was the voice of Judah, the fourth son of the clan, that was heard. Another weak link? Or a man of greater vision?

"Judah said to his brothers, "What profit is it for us to kill our brother and cover up his blood? Come and let us sell him to the Ishmaelites and not lay our hands on him, for he is our brother, our own flesh." And his brothers listened to him." (Genesis 37:26–27)

And so it was: Joseph was taken out of the cistern. Being an intelligent young man, he soon realized what could happen: they could send him far away, among the possible thousands of men in the caravan, make him disappear... What about his father? Didn't they think of their father? What would they say to him? Didn't they care about the consequences? Weren't they afraid of causing him such grief?

Joseph cried and begged his nine brothers not to harm him! His cry fell on deaf ears, empty of compassion...

They turned him over to the merchants and negotiated his price: Joseph would be sold as a slave for 20 pieces of silver and taken to Egypt, the destination of the caravan. If he were an adult, he might have been worth 30 pieces, but he was young. The brothers would have to settle for 20 pieces...

The deal was done. Joseph was tied up and taken away as a valuable item among other valuable items for which the merchants hoped to make a good profit! A distinguished, special young man, now uprooted, alone, in the midst of tough, anonymous men who spoke different languages, Joseph went along, obeying the orders he was given, sometimes only by gestures or blows that didn't visibly damage the cargo, but wounded his soul like swords. But the eyes of the Lord were still upon him.

20 coins and a big problem

Twenty coins. A small bag of 20 coins. The math was easy: 20 coins divided by 10 men equals 2 coins for each. They had made a good bargain, they thought. They had gotten rid of Joseph without killing him and made a profit of 20 coins! At last they were free of their model brother, who reproached them for simply existing, and they were richer!

But then Reuben, the eldest, arrived. He heard what had happened and was distraught! *"What am I to do?"* he asked himself and the others. It was clear to everyone that he intended to return Joseph to his father, thus perpetuating the tense atmosphere in the clan: a favorite son, with preferential treatment from his father, a dreamer, clever boy who had even received a prince's robe... But Reuben was too late! He felt a lump in his throat, a tightness in his chest. He was trapped, as if he were caught in a trap that could not be disarmed, or like Joseph himself, thrown into an empty cistern with no chance of getting out on his own... If only he could run after the caravan, bring back the money and ask them to release Joseph! No. There was nothing more to be done.

He saw no other way out but to ally himself with the nine brothers, make a deal with them, conceal the crime, and receive his two coins... With them he smeared goat's blood on Joseph's tunic. With them he gathered the flock and with them he returned to his father's house to tell him the terrible news of the disappearance of his beloved son, to lie to him and to see his great suffering.

Reuben did not denounce his brothers, and so for the next twenty-two years the ten men carried this dark secret on their consciences, pointing a finger at them day and night.

What did they do with the two coins? Did they invest in sheep? Did they buy anything they needed? They could not reveal the source of the money. Did they wrap it in a rag and bury it somewhere? We have no way of knowing, but we do know that there are profits that turn into losses: loss of peace, loss of honesty, fairness, innocence...

There are profits that steal everything that is most valuable in a person's life, leaving them alone with their inner problem of guilt, emptiness, and desire to go back and do things differently.

How was Joseph's life?

I invite you to take a look at the most important moments in Joseph's life:

- At the age of 17, in obedience to his father, he went in search of his brothers and the flocks, and was found by a man in the field who told him the direction in which his brothers had gone; he was then sold into slavery and taken to Egypt;
- He worked for Potiphar, the captain of Pharaoh's guard, and a few years later he was arrested on a false charge brought by Potiphar's wife;
- At the age of 28, he was still in prison but had become the jailer's *right–hand man (because the Lord blessed everything Joseph did)* when he interpreted the dreams of the king's cupbearer and baker;
- At the age of 30, he was summoned to the Pharaoh's court to interpret his disturbing dreams; because of his great intelligence and the solution he offered to help the country, he was made the second most important man in Egypt, governor and administrator;
- At the age of 37, Joseph saw the end of the time of abundance in Egypt and the surrounding lands, as the Lord had revealed to him through Pharaoh's dreams;
- At the age of 39, Joseph saw his ten brothers come to Egypt to buy food, fulfilling the dreams the Lord had given him as a teenager in his father's house; he finally revealed to them who he was and invited the whole family to move to Egypt because the years of famine would continue;
- Shortly thereafter, Jacob and his entire clan went to live in Egypt under the protection of Joseph, who not only returned

to embrace his father, but was with him in the last moments of his life when Joseph was 56 and Jacob was 147;

- Joseph lived to be 110 years old and saw three generations of his descendants. Throughout his life, he walked closely with God. The Lord always counted on his humility, obedience, and honesty, and in due time He fulfilled all the promises He had made to him since his youth.

What about the man in the field?

Nothing more is said about him. He was like a switch on a railroad track, also known as a needle, which allows the train to change track and direction at a certain point in its journey.

Unbeknownst to him, he was showing Joseph the way to the second most important position in the Egyptian court hierarchy, a social and political position that enabled him to escape famine and preserve the lives of his family, his people, and Egypt itself! Yes, of course, but not without first going through the experience of being threatened with death by his brothers, being sold into slavery, working as a slave, being falsely accused, becoming a prisoner among many anonymous criminals and convicts – a long period of training and discipline without which Joseph would not have developed the excellent character he came to demonstrate.

When he was released from prison, he had all the qualities, humility, and wisdom to assume the position that was always destined for him without anyone knowing except God Himself.

In his last words to his 10 sons, on his deathbed, Jacob said the following about Joseph:

"Joseph is a fruitful bough,
A fruitful bough by a spring;
Its branches run over a wall.
"The archers bitterly attacked him,
And shot at him and harassed him;

But his bow remained firm,
And his arms were agile,
From the hands of the Mighty One of Jacob
(From there is the Shepherd, the Stone of Israel),
From the God of your father who helps you,
And by the Almighty who blesses you
With blessings of heaven above,
Blessings of the deep that lies beneath,
Blessings of the breasts and of the womb.
"The blessings of your father
Have surpassed the blessings of my ancestors
Up to the utmost bound of the everlasting hills;
May they be on the head of Joseph,
And on the crown of the head of the one distinguished among his brothers.
(Genesis 49:22–26)

A moment alone with God

When I am tested, my God and Lord,
even if I don't understand why,
Make me accept the trial, not as a punishment,
but as the light touch of the God of love,
who cares for me in this way,
Who wants to make me a better man,
a son who honors his Father who is holy!
When I meet someone in the countryside,
on the road of life, at any moment,
Someone adrift, without direction or path,
Like a lost bird far from its nest,
Help me to be an outstretched hand,
Lifting up the weary, telling them they are loved
and pointing to you, my God–Savior!

A brief note on:

Children

In the Bible, Solomon's book of Proverbs presents us with texts that teach us much about family relationships, especially between fathers and their children. Let's look at the following excerpts:

"The father of the righteous will greatly rejoice,
And he who sires a wise son will be glad in him." (Proverbs 23:24)
"(...) A wise son makes a father glad (...)" (Proverbs 10:1)

All parents want their children to be wise: intelligent, reasonable, honest, diligent, obedient, grateful, fearful of God... and to have all the other qualities of excellence and rarity! This is what you can wish for your children. But a parent's expectations are not always very realistic.

Every parent tends to want a child in his or her own image and likeness. It's only natural. That's why children often hear phrases like: "At your age, I had already done...achieved...reached.... I already knew what I wanted out of life!" Etcetera, etcetera.

But now, reader, think with me for a moment: every child is created by God with unique characteristics that set him apart from everyone else in the world and from his brothers, and sisters if he has any, just as Joseph was different from his brothers, had a different calling and a special purpose in his life, planned by the Lord Himself!

A son is neither a copy nor a miniature of his father. He may or may not like the same things, have the same interests or inclinations, but he may also be almost the opposite of his father in all or some of the above! his life is no less valuable because it is different. It's just different.

Each child is a masterpiece of the Almighty! The role of the father is to raise each of his children in the knowledge of God, in the fear of the Lord, in faith (no father wants to leave behind an unsaved child when he leaves this world!), but without strangling their qualities and abilities.

As a parent, I urge you not to relinquish your role as an educator in your children's lives. Create moments for them to talk about a variety of topics beyond sports, work, church, the news—the things everyone talks about. Listen to your child's heartbeat. Get to know them. Give them opportunities to talk about what they think, what they feel, what they plan for the future.

When your children make mistakes, think about your own faults before you raise your voice beyond what is reasonable, before you get nervous or angry. See what God says about it in the Bible:

"Fathers, do not exasperate your children, so that they will not lose heart." (Colossians 3:21)

"Fathers, do not provoke your children to anger, but bring them up in the discipline and instruction of the Lord." (Ephesians 6:4)

If necessary, correct them with gentleness, as you would like them to correct you, embrace them, forgive them, encourage them, give them another chance, don't humiliate them, don't belittle them... Do with each of your children what God the Father does with us every day. Don't get tired of repeating the same things if necessary, but do it at the right time, in the right measure. You don't have to speak too loudly. It's enough to speak from the heart, with a father's love. Assure your children that you always love them, even if you don't approve or agree with their choices. In that case, explain your reasons. Children can't guess.

Don't favor one child over another, as Jacob did in his extreme love for Joseph, neglecting the others who ended up feeling rejected. Rejection opens a wound that is very difficult to treat and even more difficult to heal! All children deserve equal opportunities. Don't praise some and belittle others. Remember the things that hurt you as a child in your father's house and don't do the same.

A father's love shouldn't change like the wind, nor should his mood or the way he addresses each member of the family. This creates insecurity their children. Mey shouldn't have to check their father's face before their have the courage to talk to him about something personal. Imagine if God did that to us! How would we dare to go to him and pray? It would be very difficult...

If you are a parent, I know that you want the best for each of your children, so always respect them and you will be respected and loved. As your children grow up, support them, but with open hands, so that they can fly away and make their own way when the time is right. You don't let any of them leave home too soon to escape the heavy hand of their father. Don't want that for yourself or for them! Follow your heavenly Father as your example. In this way you will always know what to do and what to say, when to act and when not to act, when to be silent and when to speak.

May God greatly bless your life as a father and give you all the wisdom you need to deal with each of your children!

Chapter 2
Two men in prison

- Historical data -

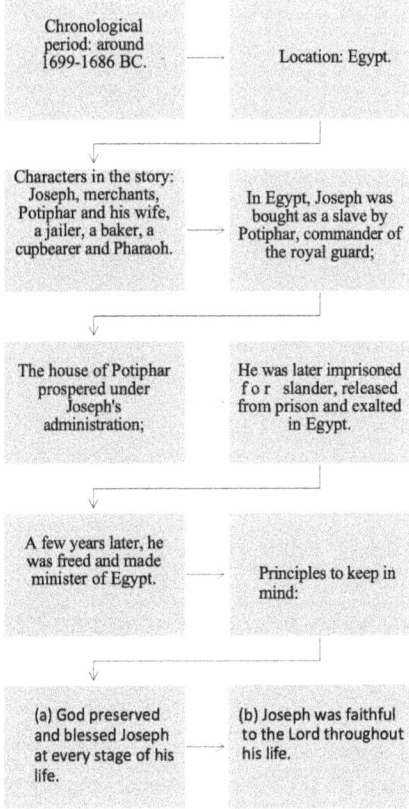

Chronological period: around 1699-1686 BC.

Location: Egypt.

Characters in the story: Joseph, merchants, Potiphar and his wife, a jailer, a baker, a cupbearer and Pharaoh.

In Egypt, Joseph was bought as a slave by Potiphar, commander of the royal guard;

The house of Potiphar prospered under Joseph's administration;

He was later imprisoned for slander, released from prison and exalted in Egypt.

A few years later, he was freed and made minister of Egypt.

Principles to keep in mind:

(a) God preserved and blessed Joseph at every stage of his life.

(b) Joseph was faithful to the Lord throughout his life.

A glimpse of ancient Egypt

In ancient Egypt, the land of pyramids, great statues and temples, prisons were part of the system of justice and re–education for citizens. They were the places where civilians who broke the law, hardened criminals, prisoners of war, enemies of the ruling regime, and anyone the judges deemed guilty were locked up. The Egyptians believed in punishment according to the crime committed.

Judges were often appointed from among men who were considered wise and capable of judging cases, who knew the laws, and also from among the priests of the temples. As a polytheistic nation (worshipping dozens of gods), the role of temples and priests was very important in society. There were prisons attached to certain temples, as the priests were the judges of the region.

3 men arrested

In the same prison, between 1690 and 1688 B.C., three men are imprisoned, accused of different crimes. Were they all guilty? Were they all innocent?

Come with me, reader, let's "enter" this Egyptian prison from more than 3700 years ago: thick walls, crude construction, little or no ventilation, a nauseating smell, very little daylight, none in some places, cells with too many men, non–existent hygiene conditions, scarce food, heat, flies and other insects, as well as crawling animals and rodents, sickness, weakness, rebellion, anger, resentment, violence, offensive language, lamentations... A cesspool of rot, torment, and bitterness, even though this prison was intended only for the king's prisoners and not for other citizens.

In a certain enclosure there are two prisoners, two court officials. The Pharaoh ordered that his chief baker and his chief cupbearer be arrested. What could have caused him to do this?

Normally, the men who held such important positions in the court, who were responsible for the food and drink brought to the royal table, were completely trusted by the king. Was there any suspicion that one (or both) of them was planning to poison the Pharaoh? Were there any reports to that effect? Was it true or false? The Bible says that they "offended their master". What kind of offense was it? Did they displease him with some attitude or word? Did they serve him food or drink that made him feel unwell?

We are not told. Until all the facts were established, until the guilt or innocence of both men was proven, the two men would remain in prison, awaiting either their release or the death penalty.

Another man approached them, the third prisoner we will now consider. He was different from the others: he showed no signs of revolt, fear, or anger. He was young, handsome, and had a pleasant presence despite the rags that covered his body. He spoke softly, trying to lift the spirits of others with words of wisdom and compassion. He seemed to occupy a prominent position there. Yes, he was indeed the right hand of the jailer. Who was this distinguished man, stuck in the same hole with the criminals, the conspirators, the prisoners of war, although he enjoyed a somewhat privileged position? It was Joseph the Hebrew, Jacob's most beloved son, his eleventh child, the boy who, at the age of 17, had been sold into slavery by his brothers and taken to Egypt.

How had he ended up in that prison? What was he guilty of? What was his crime?

Joseph had committed the "crime" of being honest and upright, correct and respectful, pure in his moral conduct. He had been harassed by the wife of his master, the captain of the guard, Potiphar, who insisted on committing adultery with this Hebrew slave. Joseph resisted all her advances and invitations until, enraged by Joseph's refusal, she planned to accuse him of trying to rape her! She accused him in sordid detail, presenting "evidence"! This alleged "evidence" was Joseph's coat, which she tore from him as he again fled her presence so as not to sin against either the Lord or Potiphar.

Apart from the garment she had torn from Joseph, there was no other evidence, only her personal testimony. When she found herself rejected again, she called several of the servants of the house and spun her tale before them: she told them that Joseph, *the Hebrew*, had attacked her at a time when all the servants of the house were away. She also accused her husband of having brought *this Hebrew* man home to mock or insult (depending on the translation of the Bible) her family!

When Potiphar returned home, she repeated the list of slanders:

"Then she spoke to him with these words, "The Hebrew slave, whom you brought to us, came in to me to make sport of me; and as I raised my voice and screamed, he left his garment beside me and fled outside."

Now when his master heard the words of his wife, which she spoke to him, saying, "This is what your slave did to me," his anger burned. So Joseph's master took him and put him into the jail, the place where the king's prisoners were confined; and he was there in the jail. But the LORD was with Joseph and extended kindness to him, and gave him favor in the sight of the chief jailer." (Genesis 39:17–21)

The woman's story was not convincing, but her husband could not allow her to be arrested for slandering the Hebrew slave! He had to choose between keeping Joseph in his service, the most reliable of the men who had worked for him, or preserving his marriage. He chose the latter.

Potiphar knew Joseph well. He knew he would never betray him, never commit adultery with his wife. He had entrusted him with the management of all the assets of his wealthy household, and nothing had ever gone missing! On the contrary, the house of Potiphar prospered under Joseph's rule. Yet he was arrested, completely innocent.

With nothing on his conscience, sure of his innocence before God and man, Joseph stood out among the prisoners: he was courteous, kind, willing to serve, humble, generous, honest, in spite of the terrible situation he was in and in spite of the fact that he had no prospect of being released!

From that dark and dirty prison, Joseph cried out to God the

Almighty, the God of Abraham, Isaac and his father Jacob. He missed his home, his family, his father, and his freedom, but that didn't stop him from being exemplary in everything he did! Within a short time, he had won the sympathy and favor of the jailer, who put him in charge of the other prisoners because the Lord was with him.

Let's look at the biblical narrative:

"So Joseph's master took him and put him into the jail, the place where the king's prisoners were confined; and he was there in the jail. But the LORD was with Joseph and extended kindness to him, and gave him favor in the sight of the chief jailer. The chief jailer committed to Joseph's charge all the prisoners who were in the jail; so that whatever was done there, he was responsible for it. The chief jailer did not supervise anything under Joseph's charge because the LORD was with him; and whatever he did, the LORD made to prosper." (Genesis 39:20–23)

God followed Joseph everywhere he went and blessed all the work of his hands!

May the Lord do so with us! May our efforts and the work of our hands prosper for His glory and the blessing of many!

First, Potiphar's house became richer; now the prison had more order, better organization, and less hatred in the air because of Joseph's work and good influence. He tried to care for the suffering of those who were imprisoned like himself, even if they were truly guilty of the crimes of which they were accused. He was sensitive to the suffering of others, perhaps because he had suffered so much in his short life...

Disturbing dreams

Have you ever had a dream or nightmare that woke you up in the middle of the night and left you restless, disturbed, and sleepless? Have you ever tried to understand or explain these strange dreams? Have you ever shared them with someone just to make sure they were nothing special? Have you ever had questions like, "Is God trying to speak to me through this dream?" Is God still speaking to people through dreams? I leave that for you to think about. Let's go back to the innocent prisoner:

When Joseph brought food to the other prisoners, he had a habit of noticing how each one looked: sad, tired, discouraged, sick... And one day he found two men who had been imprisoned for many days, troubled by the dreams they'd had the night before: Pharaoh's cupbearer and his baker. These two men had been entrusted to Joseph's care, so it was his responsibility to see that they were well, at least physically, every day. This is how the Bible tells the story:

> "Then it came about after these things, the cupbearer and the baker for the king of Egypt offended their lord, the king of Egypt. Pharaoh was furious with his two officials, the chief cupbearer and the chief baker. So he put them in confinement in the house of the captain of the bodyguard, in the jail, the same place where Joseph was imprisoned. The captain of the bodyguard put Joseph in charge of them, and he took care of them; and they were in confinement for some time. Then the cupbearer and the baker for the king of Egypt, who were confined in jail, both had a dream the same night, each man with his own dream and each dream with its own interpretation. When Joseph came to them in the morning and observed them, behold, they were dejected. He asked Pharaoh's officials who were with him in confinement in his master's house, "Why are your faces so sad today?" Then they said to him, "We have had a dream and there is no one to interpret it." Then Joseph said to them, "Do not interpretations belong to God? Tell it to me, please."

> So the chief cupbearer told his dream to Joseph, and said to him, "In my dream, behold, there was a vine in front of me; and on the vine were three branches. And as it was budding, its blossoms came out, and its clusters produced ripe grapes. Now Pharaoh's cup was in my hand; so I took the grapes and squeezed them into Pharaoh's cup, and I put the cup into Pharaoh's hand." Then Joseph said to him, "This is the interpretation of it: the three branches are three days; within three

more days Pharaoh will lift up your head and restore you to your office; and you will put Pharaoh's cup into his hand according to your former custom when you were his cupbearer. Only keep me in mind when it goes well with you, and please do me a kindness by mentioning me to Pharaoh and get me out of this house. For I was in fact kidnapped from the land of the Hebrews, and even here I have done nothing that they should have put me into the dungeon."

When the chief baker saw that he had interpreted favorably, he said to Joseph, "I also saw in my dream, and behold, there were three baskets of white bread on my head; and in the top basket there were some of all sorts of baked food for Pharaoh, and the birds were eating them out of the basket on my head." Then Joseph answered and said, "This is its interpretation: the three baskets are three days; within three more days Pharaoh will lift up your head from you and will hang you on a tree, and the birds will eat your flesh off you."

Thus it came about on the third day, which was Pharaoh's birthday, that he made a feast for all his servants; and he lifted up the head of the chief cupbearer and the head of the chief baker among his servants. He restored the chief cupbearer to his office, and he put the cup into Pharaoh's hand; but he hanged the chief baker, just as Joseph had interpreted to them. Yet the chief cupbearer did not remember Joseph, but forgot him." (Genesis 40:1–23)

God's hand was on this unique case:

1. The two men were arrested where Joseph was;
2. The two men had disturbing dreams on the same night;
3. Joseph was the only one with the gift of dream interpretation in that place, and probably in all of Egypt;
4. Joseph was responsible for the welfare of these two men.

Coincidences? No! God's wisdom. God's plan. God's specific and special purpose!

To Joseph, both dreams were as clear as daylight: they meant that the cupbearer would be restored to his position as cupbearer and that the baker would be sentenced to death by hanging. In other words, it had been proven that the cupbearer was innocent and the baker guilty. The difficulty was how to tell the two men.

It was easy to tell the cupbearer that he would return to the palace and once again serve drinks to the king, manage his cellars, supervise everything that was put on the royal table; that he would be close to the pharaoh, enjoy his sympathy, be honored, and even have the opportunity to be his advisor at times.

But how could he tell the baker that he had only a short time to live because he had been condemned? As we read in Numbers 32:23, *"Your sin will find you out."* This man's sin was going to find him sooner or later. The law of sowing and reaping has existed since the beginning of creation. The principle that *"whatever a man sows, this he will also reap"* (Galatians 6:7) has always been fulfilled and is still in effect.

Joseph was honest with his fellow prisoner. His straightforward character didn't allow him to evade the severity of the facts or sugarcoat the truth of God's revelation: the head baker would be punished for his crime. And so it was: three days later, on Pharaoh's birthday, he sent for the two prisoners and each received his sentence as Joseph had interpreted it.

A favor asked, a favor forgotten

Joseph was so sure that he had understood and interpreted the two men's dreams correctly that he asked the cupbearer for a favor even before he was called back to court and given his job back: *"Remember me if it is good for you, and I beg you to have mercy on me, and mention me to Pharaoh, and take me out of this house; for I was stolen from the land of the Hebrews, and I have done nothing here to cause them to put me in this pit."*

This was only fair. Joseph served this man during the *"many days"* he was in prison. He did his best for him. On the day he saw him at his worst, troubled by his dream, he patiently listened to his story and, with the wisdom God gave him, interpreted his dream and set his mind at rest.

The cupbearer, however, when he found himself back in his position of honor, put Joseph's plea to the back of his mind... His problem was already solved, he didn't need to get involved in other people's affairs, let alone a Hebrew slave in prison! That was none of his business! Just imagine: Pharaoh might not take kindly to this sort of intercession on behalf of the jailer's slave. And was he really as innocent as he claimed? And what if he wasn't? The cupbearer didn't want to jeopardize his position again... And so nothing changed: Joseph remained imprisoned, innocent, in that tomb... Not because of the "forgetfulness" of the cupbearer, but because the Lord had a purpose: Joseph was still in the hands of the potter, being molded with all precision, being prepared for the time to come.

Reader, when it seems to you that everyone has forgotten you, including God, reject that thought! When the Lord seems to ignore your prayers and supplications, your efforts and struggles, believe this:

- *The love that the Lord has for you is unquestionable, unwavering, and infinite, proven on the cross of Calvary;*
- *He is preparing the blessing He is going to give you, or preparing you to receive it; if the Lord's final answer to your prayer is ultimately "no," still believe that His thoughts are higher than yours; we don't understand everything that happens to us on this side of life, but we will in eternity;*
- *He is transforming your life, shaping your character, perfecting some trait of Christ in you;*
- *He is transforming you into the blessing you will become for someone, for many or for a few, it doesn't matter! What matters is that you exist to honor and glorify the Lord!*
- *He will never abandon you: he has promised to be with you in every circumstance, every day, and he will keep that promise in full, because the Lord is faithful!*
- *Be strong and courageous in adversity, persevering in faith and patient in tribulation (James 1:3; 1 Peter 1:7).*

God's intervention

Joseph's appeal to the king's butler was forgotten. So it seemed. The biblical account of this event leads us to believe that it was intentional: *"Yet the chief cupbearer did not remember Joseph, but forgot him."* *(Genesis 40:23)*

But God remembers the good we do and our prayers, which are always before Him. His plan for Joseph went far beyond anything the head butler could do for him, and far beyond anything Joseph could imagine! His cause might have been forgotten by all, but not by the Lord! More time was needed for Joseph to get to the right place, to get the temperament needed for the place God had chosen for him. But Joseph didn't know that...

After the release and restoration of the head butler, Joseph waited for someone to call him from prison with the long–awaited news: "You're free! It has been confirmed that you are innocent!" But the days passed with nothing happening. They gave way to months of waiting and silence. The cupbearer had forgotten him... Or he hadn't found the right opportunity to speak to the king... Or his request had been refused... A whole year passed, and then another year... God was still with Joseph in prison. He continued to refine his qualities, slowly and deeply.

Then suddenly the prison receives orders to release Joseph and bring him before Pharaoh – now! But why? Did the cupbearer intercede on his behalf? No, but the Lord made him remember Joseph and see his carelessness and ingratitude all at once:

One night God brought the king disturbing dreams, dreams that robbed him of sleep, that left him perplexed, restless, and anxious to know what they meant. When the wise men, soothsayers and counselors gathered, they all admitted that they didn't understand the dreams, let alone what they meant. Then, yes, it was time for God to intervene on Joseph's behalf through the voice of the cupbearer. The Lord jogged his memory and his thoughts flew back to when he had been in prison and had met an extraordinary young man who could interpret dreams!

"Then the chief cupbearer spoke to Pharaoh, saying, "I would make mention today of my own offenses. Pharaoh was furious with his servants, and he put me in confinement in the house of the captain of the bodyguard, both me and the chief baker. We had a dream on the same night, he and I; each of us dreamed according to the interpretation of his own dream. Now a Hebrew youth was with us there, a servant of the captain of the bodyguard, and we related them to him, and he interpreted our dreams for us. To each one he interpreted according to his own dream. And just as he interpreted for us, so it happened; he restored me in my office, but he hanged him."

Then Pharaoh sent and called for Joseph, and they hurriedly brought him out of the dungeon; and when he had shaved himself and changed his clothes, he came to Pharaoh." (Genesis 41:9–14)

This text shows us how important the role of the head butler was at the king's court to the point that Pharaoh accepted his suggestion because Joseph's credentials as a dream interpreter had been verified by the confirmation of his words in the lives of the butler and the baker.

Soon, everything changed for Joseph. After interpreting the king's dreams, he also advised him on how to avoid famine and scarcity in Egypt.

"Pharaoh said to Joseph, "I have had a dream, but no one can interpret it; and I have heard it said about you, that when you hear a dream you can interpret it." Joseph then answered Pharaoh, saying, "It is not in me; God will give Pharaoh a favorable answer."" (Genesis 41:15–16)

This young man, just released from prison after four long years of unjust imprisonment, now dressed in clean clothes and clean-shaven, was very different from the boy who had dreams in his father Jacob's house and told them to his family with a certain smugness.

Over the years, he had been crushed by pain on many levels, that pain that wounds the human soul to the core. He had experienced the envy of his brothers, separation from his father, slavery, shame, slander, unjust imprisonment... A lot of humiliation, a lot of abandonment, a lot of tears that no one saw but God. And not only did he survive it all, but he came out of the circle a better man. He was tested in God's crucible and came out like refined gold. This qualified him to become the second man in charge of Egypt, a great and powerful nation at that time.

Reader, I don't know your story. I don't know what wounds have scarred your character over the years, I don't know about your sorrows or your struggles. I don't know how you feel, nor can I imagine it. We're all so different, each with our own story, our own path...

But I do know that God sees, understands, and never crushes the broken reed or the smoldering wick. What's more, He has the power to heal wounds and restore broken vessels. He has seen you in every situation: happy, strong, enthusiastic, full of plans and extraordinary ideas, applauded and loved! And He has also seen you sad, defeated, without ideas, without plans, alone, disoriented, weak and sick in body and soul. Nothing is hidden from Him. Nothing has happened to you that He doesn't know about. You can tell Him everything and talk to Him about everything that still hurts because He understands and is ready to embrace you like no one has ever embraced you before.

The king's conclusion

After listening to Joseph's interpretation of his dreams, the king concluded that he was the right man for this moment in Egypt's history. He had a different kind of wisdom from God. So he decided to give him the enormous task of organizing the nation's economy and managing its agricultural production.

"So Pharaoh said to Joseph, 'Since God has informed you of all this, there is no one so discerning and wise as you are. You shall be over my house, and according to your command all my people shall do homage; only in the throne I will be greater than you.' Pharaoh said to

Joseph, 'See, I have set you over all the land of Egypt.' Then Pharaoh took off his signet ring from his hand and put it on Joseph's hand, and clothed him in garments of fine linen and put the gold necklace around his neck. He had him ride in his second chariot; and they proclaimed before him, "Bow the knee!" And he set him over all the land of Egypt. Moreover, Pharaoh said to Joseph, 'Though I am Pharaoh, yet without your permission no one shall raise his hand or foot in all the land of Egypt.'" (Genesis 41:39–44)

The rest of the events related to Joseph are recorded in the book of Genesis, from chapter 41 to chapter 50.

The Word of God does not mention the name of Pharaoh's chief butler, who played a decisive role in Joseph's exaltation and, indirectly, in the survival of the Egyptians and the people of the surrounding region, who were struck by famine and turned to Egypt for food. If he hadn't mentioned the name of Joseph, the Hebrew servant, and his qualities before the court and Pharaoh, the whole history of that region would have taken a different course, not only Joseph's life, but that of thousands and thousands of families.

One last word, reader: always try to do the right thing and don't worry about your name remaining anonymous. God knows. That's enough. Not even a glass of water given as a disciple goes unrewarded.

"And whoever in the name of a disciple gives to one of these little ones even a cup of cold water to drink, truly I say to you, he shall not lose his reward." (Matthew 10:42)

A moment alone with God

Prayer:

Lord, when I feel wronged, slandered, mistreated, and cruelly treated, help me to remember the Lord Jesus Christ, who for my sake suffered what I have never suffered and will never suffer. My pain cannot compare to His who gave Himself for me while I was still a sinner!

Lord, I pray that when I am tempted to "pay in kind," You will remind me of the supreme example of my Savior, who "when he was reviled, did not return the insult; when he was abused, did not threaten, but committed himself to Him who judges righteously" (1 Peter 2:23).

On the other hand, help me to be grateful in all circumstances and to repay, to the best of my ability, all the good that has been done to me. Help me to be the best person I can be – by Your grace and for Your glory!

Help me to learn to "lose" my life by spending it in your service, knowing that I will receive a servant's reward if I know how to be good and faithful (Matthew 25:21). Whatever comes into my hands to do, give me the ability to do it with all my might (Ecclesiastes 9:10), because it is for You and not for ordinary people like me (Colossians 3:23).

And, Lord, if You give me the opportunity to show someone the way, give me the wisdom, courage, and love to do it faithfully. I pray in the name of the Lord Jesus Christ.

Chapter 3
A boy and a hero

- Historical data -

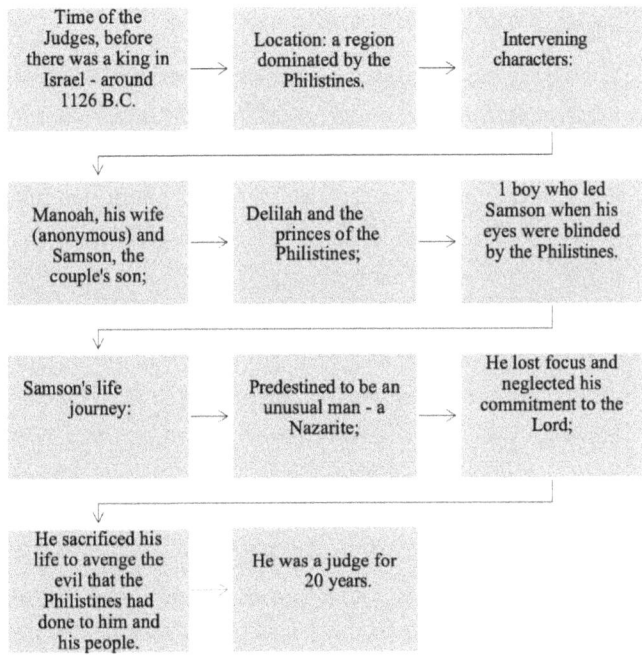

Time of the Judges, before there was a king in Israel - around 1126 B.C.	Location: a region dominated by the Philistines.	Intervening characters:
Manoah, his wife (anonymous) and Samson, the couple's son;	Delilah and the princes of the Philistines;	1 boy who led Samson when his eyes were blinded by the Philistines.
Samson's life journey:	Predestined to be an unusual man - a Nazarite;	He lost focus and neglected his commitment to the Lord;
He sacrificed his life to avenge the evil that the Philistines had done to him and his people.	He was a judge for 20 years.	

A careless hero

Scripture tells the story of a childless couple, Manoah and his wife.

God chose them to be the parents of a special boy, endowed with extraordinary strength, who in time would be a powerful weapon in God's hands to liberate the Israelites from Philistine domination. They named him Samson. This story is found in the book of Judges, chapters 13 through 16.

Samson soon became a temperamental young man. There were three unquestionable rules that were to guide his life, to serve as boundaries that would keep him in complete safety: not to drink alcohol or even eat grapes, not to touch dead bodies, and not to cut his hair. These were the 3 signs of his Nazaritism, of his exclusive consecration to God. He was *"a Nazirite to God from the womb to the day of his death."* (Judges 13:7)

Chosen by God for a mission even before he was born, this young man relied on his physical strength and was easily led by his passions. This way of life brought him countless sorrows, despite the many victories he also achieved over the oppressive people of Israel.

Eventually, he fell in love with Delilah, a Philistine woman who loved money more than Samson himself. She was the person he went to when he needed to rest. She would let him rest his head on her knees and listen to him tell her of his achievements. Until some Philistine princes became aware of Samson's habit of visiting Delilah's house and thought that she would be the ideal person to find out what his supernatural strength came from and hand him over to them as a weak captive. Delilah agreed to the proposal, and they agreed on the price she would receive for betraying Samson.

Unaware of all this, he continued to visit his beloved to confide in her everything that was deep inside him, except the source of his strength! With patience, persistence, skill, and emotional blackmail, Delilah finally managed to learn Samson's great secret: if his hair were cut, his supernatural strength would disappear and he would be like any other man...

What a mistake, Samson! Can't you see she's a Philistine?! There were so many things you never told your father or your mother, and now you're letting your guard down and opening your heart completely to a foreign woman who you think loves you, who you think you know, but you're so wrong...

So while Samson slept on Delilah's lap, his hair was shaved off... When he woke up from his sleep, awakened by the commotion of the Philistines ready to arrest him, he discovered that his head was bare, without his seven long braids, and his strength was gone: he was just an ordinary man – betrayed by the woman he loved...

Samson in prison

What a sad story! What a tragic end! Bound like a defeated beast, Samson became "a grinder in the prison" (Judges 16:21) – like a beast of burden.

If Satan can find a loophole in your life, as he did in Samson's, he will do everything in his power to destroy everything he can touch: your job, your ministry, your fellowship with God, the unity of your family, your greatest love, your passion for noble causes, your good name – everything, everything he can! Remember, he "prowls around like a roaring lion, seeking someone to devour" (1 Peter 5:8). Be careful! Don't let your guard down! Live according to God's holy and righteous principles! "Watch over your heart with all diligence, for from it flow the springs of life." (Proverbs 4:23)

But God has a gentle and powerful way of giving us – all of us – a second chance! The Lord intervened on Samson's behalf: He gave him the opportunity to restart his life of faith, to restore his fellowship with God. Look at what the Word says in Judges 16:22:

"[T]he hair of his head began to grow again after it was shaved off."

To the Lord be all glory! He is the God who does not forsake us, who carries out to the end the good work begun in each one of us, even if it is necessary to allow us to come to the end of our own strength or to be reduced to a position much lower than the one He had prepared for us. If this is what it takes for us to truly repent of our sins and return to the open arms of our Heavenly Father, He will allow it!

A small detail:

Samson was blind, his eyes had been gouged out by his enemies, so he was given a boy to guide him. This small detail, so simple, so seemingly natural, was the lever that led Samson to ultimate victory!

A boy and a hero at the big feast

BIBLICAL NARRATIVE

Thus says the Word of the Lord:

> *Then the Philistines seized him and gouged out his eyes; and they brought him down to Gaza and bound him with bronze chains, and he was a grinder in the prison. However, the hair of his head began to grow again after it was shaved off.*
>
> *Now the lords of the Philistines assembled to offer a great sacrifice to Dagon their god, and to rejoice, for they said,*
> *"Our god has given Samson our enemy into our hands."*
> *When the people saw him, they praised their god, for they said,*
> *"Our god has given our enemy into our hands,*
> *Even the destroyer of our country,*
> *Who has slain many of us."*
>
> *It so happened when they were in high spirits, that they said, "Call for Samson, that he may amuse us." So they called for Samson from the prison, and he entertained them. And they made him stand between the pillars. Then Samson said to the boy who was holding his hand, "Let me feel the pillars on which the house rests, that I may lean against them." Now the house was full of men and women, and all the lords of the Philistines were there. And about 3,000 men and women were on the roof looking on while Samson was amusing them.*
>
> *Then Samson called to the LORD and said, "O Lord GOD, please remember me and please strengthen me just this time, O God, that I may at once be avenged of the Philistines for my two eyes." Samson*

grasped the two middle pillars on which the house rested, and braced himself against them, the one with his right hand and the other with his left. And Samson said, "Let me die with the Philistines!" And he bent with all his might so that the house fell on the lords and all the people who were in it. So the dead whom he killed at his death were more than those whom he killed in his life. (Judges 16:21–30)

Samson was brought to the feast of his enemies to amuse them, to humiliate him even more than he had already been humiliated, and to thank their gods, namely Dagon (represented by an image that was half man and half fish), for the victory over their greatest nightmare: God's Nazirite, the man born to be the judge of his people, to defend them, and to bring peace to the nation.

This ancient Israelite hero was now reduced to a weak, stumbling, blind man who had to be guided by the hand of a boy...

During the time that Samson was in prison turning a mill, not only did his hair grow back, but the muscles in his arms grew stronger. His enemies didn't notice this transformation, but God saw him in his daily efforts to turn the millstone and grind the grain, and each day he added the qualities he needed to win his ultimate victory.

On the day of the feast, in the temple of Dagon, the ancient hero of Israel entered the great hall, led by the hand of the boy. We don't know his name. But we do know that he led Samson to the place he had asked for: the center of the great house, where there were two pillars:

"Then Samson said to the boy who was holding his hand, "Let me feel the pillars on which the house rests, that I may lean against them.""
(Judges 16:26)

We would have liked to read that Samson told his young servant to get out of the place quickly! But this does not appear in the Scripture. Whether the young man understood what was about to happen, and whether he had time to flee, we do not know. The Lord chose not to tell us the boy's name or what happened to him. He was an anonymous young man whom God used at the right time to put

Samson in the right place in that idolatrous temple.

Nothing was expected there but amusement for the Philistines and humiliation for the blind and defeated hero. Samson was not expected to say his last prayer in that place! The Lord was not expected to answer his plea as he did! It wasn't expected that one man alone would have the strength to bring down such a great building! And yet, all this happened because Samson's God is our God, the Lord of hosts, the Almighty!

Without sight, without a sword in his hand, Samson was God's secret weapon at that time. He sacrificed his own life and carried out his mission selflessly. It didn't have to be this way if he had lived his life according to God's will.

Reader, I would like to leave you with two notes at the end of this chapter:

1. *There are no inappropriate places to pray. Any place, at any time, is the right place to cry out for help from the Lord, the One who can do anything, who transforms minds and hearts, who gives peace, courage and victory in the most unexpected moments, in the most unlikely places;*

2. *No child of God can exhaust the Heavenly Father's love, nor his superabundant grace, nor his compassion: the Lord is the God of second chances, forgiveness, restoration, and a fresh start in faith; all you have to do is cry out to him.*

A moment alone with God

Prayer

I know that You have made me for a purpose, and I also know that I have not always been on the path You have set for me.

Today I come to renew my commitment to You and by faith I receive a new opportunity to be better this time. In the name of the Lord Jesus Christ.

An anonymous man in prison

In the dark prison, turning a millstone,
Grinding the hard grain, abandoned and alone,
I see a hero, once proud and strong,
Now broken, bound and captive,
With no one to cheer or comfort him.
His only companion is a servant,
His lookout, help and only support,
The men who took him there think...

They can't imagine the invisible God,
In silence, incognito, in intangible presence,
Is also inside that prison,
With his eyes resting
On the slumped shoulders of a defeated Samson.
But nothing is lost: the Lord is there,
He renews his faith, gives strength to his arms,
Keeps him on his feet for the last blow!

In the center of the house, between the two pillars,
Samson hears the laughter of mockery and derision.
From his soul comes a cry worth thousands of cries!
I don't know if he weeps, but I do know that he prays
To his God and Lord!
He asks for the strength to win one last time,
Without pride, vanity or any haughtiness.
God hears and answers his prayer:
He finally gives victory to his servant Samson!

To the Lord alone I give all the glory!
To the God who moves and acts on my behalf:
He is the God of Samson, the God of Abraham,
Of infinite power, infinite forgiveness and infinite love!

Chapter 4
Thirty men and a friend

- Historical data -

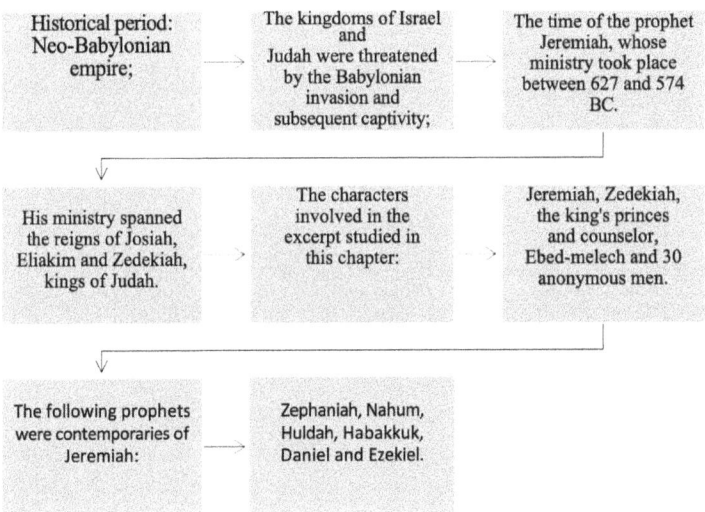

Historical period: Neo-Babylonian empire;	The kingdoms of Israel and Judah were threatened by the Babylonian invasion and subsequent captivity;	The time of the prophet Jeremiah, whose ministry took place between 627 and 574 BC.
His ministry spanned the reigns of Josiah, Eliakim and Zedekiah, kings of Judah.	The characters involved in the excerpt studied in this chapter:	Jeremiah, Zedekiah, the king's princes and counselor, Ebed-melech and 30 anonymous men.
The following prophets were contemporaries of Jeremiah:	Zephaniah, Nahum, Huldah, Habakkuk, Daniel and Ezekiel.	

In the biblical text we are about to read, recorded in the book of the prophet Jeremiah, we encounter various characters that we will consider, along with additional information that is important for understanding the period.

Various places are also mentioned, notably the *cistern of Malchijah, which was located in the courtyard of the guard.*

In a land on the brink of war, with invaders approaching the gates of the city of Jerusalem, a voice rises that no one can silence: it is the voice of Jeremiah, the prophet of God. He faithfully warns the people what to do if the invaders take the city. The words that God gives him, he passes on to the people of Judah, leaving nothing out. There are many discontented people, especially among the upper classes, who always have the most to lose.

Parties formed. King Zedekiah was confused, he didn't know who to listen to. On the one hand, the king himself threw the parchment into the fire.

On the other hand, he sent secret messages to the prophet of God, saying: *"Please pray to the LORD our God on our behalf."* (Jeremiah 37:3). A strong movement arose against Jeremiah's message and he was arrested.

Well or cistern?

Several passages in the Bible refer to wells, cisterns, tanks, dams, aqueducts, and fountains. These were different ways of collecting, storing, and disposing of water, such a precious and sometimes scarce commodity in that geographical area. Here we'll just see that wells and cisterns are not exactly the same thing.

According to the Archaeological Study Bible (NIV) by Editora Vida, a cistern was a type of reservoir for collecting rainwater. It was usually shaped like a bottle or a bell. It had a narrow opening at the top to prevent evaporation. The inside of the cisterns was lined with an insulating material so that all the water collected was preserved and not lost. Drawing water from a cistern was the same as drawing water from a well.

In biblical times, homes had private cisterns, but cities also had larger public cisterns. Archaeological excavations north of Jerusalem have revealed very large cisterns, the largest of which can hold 9,000 cubic meters of water. It is known as the Great Sea and is 13 meters deep.

A well was – and still is – *a pit or hole sunk into the earth to reach a supply of water (according to Merriam–Webster)*. It is dug in impermeable stone and plastered to prevent the infiltration of less clean water.

In this chapter, we will discuss the courageous act of thirty men who joined together for a rescue led by Ebed–melech, an Ethiopian in the service of King Zedekiah of Judah.

These thirty–one men formed the rescue team that managed to save the prophet Jeremiah from the death that awaited him at the bottom of a cistern that no longer contained drinking water, but only mud.

How could a prophet of God fall into a cistern?

Or was he pushed? By whom? For what purpose? What an extraordinary case!

Come with me and let's have a look at the bottom of this cistern...

A friend at the bottom of the cistern and a friend outside the cistern

BIBLICAL NARRATIVE

"Now Shephatiah the son of Mattan, and Gedaliah the son of Pashhur, and Jucal the son of Shelemiah, and Pashhur the son of Malchijah heard the words that Jeremiah was speaking to all the people, saying, "Thus says the LORD, 'He who stays in this city will die by the sword and by famine and by pestilence, but he who goes out to the Chaldeans will live and have his own life as booty and stay alive.' Thus says the LORD, 'This city will certainly be given into the hand of the army of the king of Babylon and he will capture it.'" Then the officials said to the king, "Now let this man be put to death, inasmuch as he is discouraging the men of war who are left in this city and all the people, by speaking such words to them; for this man is not seeking the well–being of this people but rather their harm." So King Zedekiah said, "Behold, he is in your hands; for the king can do nothing against you." Then they took Jeremiah and cast him into the cistern of Malchijah the king's son,

*which was in the court of the guardhouse; and they let Jeremiah down
with ropes. Now in the cistern there was no water but only mud, and
Jeremiah sank into the mud." (Jeremiah 38:1–6)*

Look, reader! There he is, our man, the prophet Jeremiah! As we
read the biblical text, we can easily understand why he was at the
bottom of the cistern: because he was faithfully proclaiming God's
message to the nation of Israel.

God's message to man is not always to his liking. It often goes
in the opposite direction from what everyone wants to hear. In this
case, it was a harsh message, telling the people to surrender to a
foreign nation instead of trying to stand up and resist.

King Zedekiah was confused. He reigned in Judah from 597 to
586 B.C. His advisors took a firm and uncompromising stand against
God's message delivered through the prophet Jeremiah. They couldn't
silence the man! They had to take more drastic measures!

*Reader, if God gives you a message to deliver to the people, honor Him
with the obedience of faith. Preach, proclaim or write that message clearly
so that even those who are rushing through life will have the opportunity to
read or hear it! Don't give up! It's a mission from the Lord! See it through!
And leave the rest to Him, whatever you can't solve or change by your
own strength or abilities.*

Jeremiah did not remain silent. He couldn't suppress or ignore
God's word to his people. He knew the consequences of preaching his
message. He knew he would face persecution and serious problems,
even death, but he remained faithful to the Lord.

Yes, the king's advisors wanted to kill him, to silence him forever!
Zedekiah was in a position of weakness and minority. He gave in.
He delivered the prophet into the hands of his enemies.

His fate was quickly decided: he would be thrown into the nearest
cistern, Malchiah's, which was in the prison yard, and he would die in
the mud! It was a deep cistern. They lowered him down with ropes.

As we saw earlier, the opening of the cisterns was small and the
space widened according to the depth. Look now, reader: there is

Jeremiah, the servant of God, stuck in the mud at the bottom of the cistern. He is only visible when daylight shines on him. When it gets dark, no one will see him. No one will bring him water or bread. He will be forgotten. When he gets tired of standing, when his strength runs out and he falls asleep from extreme exhaustion, he will fall into the mud and that will be the end of his life... The cistern will be his grave...

No, this is not the end of the story! The prophet Jeremiah had an unlikely friend outside the cistern who had seen what they had done to him, who knew where he was and the risk he was taking. The man's name was Ebed–melech, which means *servant of the king.*

Read it with me, please:

But Ebed–melech the Ethiopian, a eunuch, while he was in the king's palace, heard that they had put Jeremiah into the cistern. Now the king was sitting in the Gate of Benjamin; and Ebed–melech went out from the king's palace and spoke to the king, saying, "My lord the king, these men have acted wickedly in all that they have done to Jeremiah the prophet whom they have cast into the cistern; and he will die right where he is because of the famine, for there is no more bread in the city." Then the king commanded Ebed–melech the Ethiopian, saying, "Take thirty men from here under your authority and bring up Jeremiah the prophet from the cistern before he dies." So Ebed–melech took the men under his authority and went into the king's palace to a place beneath the storeroom and took from there worn–out clothes and worn–out rags and let them down by ropes into the cistern to Jeremiah. Then Ebed–melech the Ethiopian said to Jeremiah, "Now put these worn–out clothes and rags under your armpits under the ropes"; and Jeremiah did so. So they pulled Jeremiah up with the ropes and lifted him out of the cistern, and Jeremiah stayed in the court of the guardhouse. (Jeremiah 38:7–13)

Ebed–melech was the unlikely friend outside the cistern. Unlikely, because he wasn't even Jewish. He was an Ethiopian. Nor was he a prince or an advisor to the king. He was a servant. He wasn't even a man like all the others. He was a eunuch.

How God uses the weak things of this world to confound and shame the strong! (1 Corinthians 1:27) Praise be to His name!

Ebed–melech genuinely cared for the prophet Jeremiah like a true friend. His background, profession, social status or health were not important when a man's life was at stake! Once he realized the situation, he acted boldly and wisely, with the utmost compassion:

- He looked for a moment to speak to the king alone;
- He explained the deplorable conditions in which Jeremiah found himself;
- He warned him of the high probability that the prophet would lose his life in that mud hole.

Ebed–melek was a kind of Good Samaritan before the time when this teaching would be brought by the Lord Jesus Christ.

He didn't think of himself. He wasn't afraid that by coming to Jeremiah's defense, he might be thrown into the same cistern! His boldness and selflessness were rewarded when the king ordered him to organize a group of 30 men and pull Jeremiah out of the mud pit.

Reader, when the Lord gives you a vision of ministry or shows you a need in your workplace, your church, your family, or anywhere else in the world, He will probably send you to meet that need. That doesn't mean He will send you alone, but you will certainly be part of the team on that mission!

Now Ebed–melech would have to gather 30 men to form the prophet's rescue team. It couldn't have been easy. He was inviting men to a difficult mission, not to a party! Perhaps not everyone was eager to accept. But the king's servant was not discouraged. He continued to recruit men until he had his team of 30, as the king had ordered. He didn't turn around and complain about some who didn't want to go, or resign his commission and say that the king had better give the orders himself... None of that! It was his responsibility. It was his life's mission!

Nor did he think that 10 or 12 men would be enough. He probably had no experience in rescuing people stuck in the mud at the bottom of a cistern… But had he dealt with anything like this in his life? We don't know, but we wouldn't be surprised if he had.

Once the men were assembled, and before they went to the rescue site, Ebed–melech arranged for ropes and went to get old clothes that could be used to cover Jeremiah's armpits as he was lifted out of the cistern. Who would remember this detail except God, our merciful and compassionate God, or someone who had been in a similar situation?

Ebed–melech's past is not revealed to us in the Scriptures. We know nothing of his life experiences from his birth until he arrived in Judah and became the king's servant. However, his understanding of human frailty, his understanding of hunger and suffering in general, and his compassion lead us to believe that this man had a painful history.

30 men

Thirty men volunteered to help another man who was at the bottom of a muddy cistern. They put their strength, dexterity, and skill to the service of others under the leadership of Ebed–melech. He wasn't an elite warrior that everyone could trust or rely on in risky situations. He didn't have a degree in anything. He hadn't received any preparation or special training to help with rescues. He wasn't one of the king's sons. He didn't own or control anything except his integrity and his compassion. Yet there he was on the front line, the only man who cared, who bothered to do something to save Jeremiah from the death that awaited him, the only man who wanted to make a difference in someone's life. This courageous attitude inspired 30 other men! Blessed contagion! With what passion did he speak of the prophet? One day we'll find out…

Thirty men planned the rescue. Thirty men, whose names are not recorded in the Bible but who are certainly remembered before the Lord, found the most effective way to rescue Jeremiah.

There were difficulties: the opening of the cistern was not large;

Jeremiah was tired, dehydrated, lacking in strength, probably injured, and unable to cooperate much... It was necessary to find something strong to which the ropes could be tied and which would act as a fulcrum, a kind of rudimentary pulley. This "strong thing" might have been a group of men among the 30, whose job was to stand firm, as if rooted to the ground, while the rest of the group organized itself.

Each took his place according to his ability and strength. Ebed–melech explained the plan to the prophet, threw him the old clothes and ropes, and instructed him how to attach himself to the ropes so that he could be pulled up, which he did.

Then the men pulled together, slowly, without stopping, until Jeremiah emerged from the cistern, dirty, full of mud, but alive, as alive as the message he never got tired of repeating! How could he thank all those men and Ebed–melech? There were no words. God had given him a brave and compassionate friend who had managed to gather 30 equally brave and compassionate friends to pull him out of the mud of the cistern and bring him to safety. God is infinitely good and always knows where we are!

If you happen to have a friend at the bottom of a pit, why not think about putting together a rescue team and throwing them a strong rope?

The reward

Jeremiah remained a prisoner, but now in the courtyard of the guard, where he received water and bread every day until the prophecy was fulfilled: Jerusalem was attacked and taken. Jeremiah survived, and the Lord gave him a special message for Ebed–melech:

> *"Go and speak to Ebed–melech the Ethiopian, saying, 'Thus says the LORD of hosts, the God of Israel, 'Behold, I am about to bring My words on this city for disaster and not for prosperity; and they will take place before you on that day. But I will deliver you on that day,' declares the LORD, 'and you will not be given into the hand of the men whom you dread. For I will certainly rescue you, and you will not fall*

by the sword; but you will have your own life as booty, because you have trusted in Me,' declares the LORD.'" (Jeremiah 39:16–18)

More than a servant of the king, Ebed–melech was a servant of God who was able to motivate 30 men to participate in the rescue of the prophet Jeremiah. He was an encourager. The world needs people like that!

If you are asked to be part of a rescue team, a prayer group, or any other kind of intervention for someone who is in a mud cistern, in a pit of any kind, do not hesitate! Be like one of the 30 anonymous people who made all the difference in the life of an innocent man, a messenger of God and a nation!

The pits and cisterns of life

We're now going to refer to difficult situations in life and compare each one to a deep pit, since pits are more common these days.

Regardless of our social class, academic degree, professional status, politeness, emotional balance, or any other apparent advantage, we are all at risk of falling or being thrown into the bottom of a pit. We're not talking about water pit, but emotional pits, spiritual pits, and many of life's circumstances that resemble deep, dark pits that we think we'll never get out of, that will become a permanent part of our history, that will never allow us to get up, climb out, and walk back into the light of day – out of the pit – with our heads held high.

There are times in life when we "fall" slowly into the abyss, as if we were watching a movie in very slow motion... We fall without realizing that we're descending to a dark, lonely bottom with no way out. There are no exits at the bottom! The only way out is up, from where we're slowly moving away, a little more with each passing day.

Some pits have support points, small iron bars attached to the wall of the pit, like a vertical ladder, to allow descent and ascent.

Others simply have smooth, circular walls extending all the way to the bottom, with no way to climb out.

Some situations throw us to the bottom of a pit with no time to escape. The Bible tells of such a case: ten men, out of jealousy, threw his brother Joseph down a pit, despite his pleas. They then pulled him up to sell him as a slave to a caravan of traders on their way to Egypt. Joseph was the favorite son of Jacob, the son of Isaac, the son of Abraham, the founder of the nation of Israel. Joseph was the apple of his father's eye. The trustworthy son. The obedient son. The son every father would want to have.

His brothers were different, headstrong, impulsive, careless, resentful, and as a result they couldn't handle their father's preference for Joseph. Then, one day, an opportunity arose to take revenge, to "teach their father's boy a good lesson"... Brothers throwing brothers into the deep end...a story like so many stories today, not just an incident from ancient times. (See Part IV, Chapter 1)

Jesus' brothers and the Pharisees, who considered Him an enemy, also tried to do the same – to throw Him into the pit of fame, where they hoped He would get stuck in the mud of pride and never come up again to teach the truth to the people; to push Him into pits of dilemmas, difficult choices, complicated questions, to trap Him, to close off all possible exits... Satan tried the same thing, in many ways, with many tricks. He tried until the last moment, even when Jesus was nailed to the cross, suggesting that He climb down... But of course no one succeeded, because the Lord was not like His brothers, nor like the Pharisees, nor like any tempter. He was Holy God made man, so no ordinary man or host of evil could do anything against Him without His lawn permission!

In the case of the prophet Jeremiah, he was thrown into the bottom of a mud pit by the authorities of his people, the princes and advisors of King Zedekiah, who reigned in Judah for eleven years, between 597 and 586 B.C. These authorities, the prophet's fellow countrymen, did not want to hear the voice of God. Jeremiah, however, was a man with a mission: to deliver the word of the Lord to his people as he had given it to him. To warn them of the impending invasion of Jerusalem by the Chaldeans and the consequences that would follow if the people resisted the invaders.

We tend to ignore or place little value on people who speak to us in the name of the Lord. Sometimes we think, "How can he claim to have heard God's voice when I've never heard anything? Is God speaking to him and not to me?" You know, reader, it's possible. The only question is why.

Jeremiah walked with God in such a way that they became close friends. The Lord didn't have to shout for the prophet to hear him. All he had to do was whisper in his ear the message that urgently needed to be delivered to the people, and Jeremiah would rise to obey. Let us do the same!

Pits where God meets us

The Lord finds us and meets us in the deepest, darkest, dirtiest pits, and He pulls us out! He doesn't want us to stay there for long, just long enough to learn the lesson we missed. Sometimes through friends, sometimes through anonymous people, sometimes through the power of His hand stretched out in our favor, or through the power of His Word, the Lord makes us rise from the depths and face the light of day with new courage and new knowledge.

Would you like to have a look at some of these pits?let's

- The pit of sin;
- The pit of rebellion, stubbornness and pride;
- The pit of discouragement;
- The pit of self–sufficiency;
- The pit of a busy schedule;
- The pit of pride and vanity;
- The pit of fear;
- The pit of discouragement and despondency;
- The pit of loneliness;
- The pit of sickness and physical pain;
- The pit of depression, emotional and spiritual pain;
- The pit of anger and rage;
- The pit of indecision and procrastination;

- The pit of disorder and chaos;
- The pit of laziness;
- The pit of pretense and hypocrisy;
- The pit of debt;
- The pit of unforgiveness;
- The pit of envy;
- The pit of the desire for revenge;
- The pit of the past;
- The pit of self–pity and low self–esteem;
- The pit of selfishness;
- The pit of addiction;
- The pit of indifference.

As you looked at this list, you may have thought of other pits that you know of, that you've seen, that you've helped someone get out of, or that you've gotten out of by the grace of God. You can add them to the list if you wish.

The pit as a meeting place

There are times in life when we are struggling at the bottom of some kind of pit. We long for the outstretched hand of someone to help us up. Even a rope would give us hope, let alone a firm hand! A hand capable of reaching us, capable of reaching down far enough to grab us, strong enough to pull us out and give us a chance to see things in the full light of day. It would be so special if someone could throw us a strong rope and maybe some rags to cushion the pain of the rope that would pull us...

There are other times in life when we've been at the bottom of the pit for so long that we've even developed strategies to continue living in that dark and difficult environment. We almost make the pit our home, the deep hole our home. We almost don't want to leave. We no longer believe in any other way of life. We don't want to be bothered, to be invited out of the pit, to be given a rope. We don't trust anything new or different. We get used to the pit and that's it!

The pit of depression. The pit of self–sufficiency. The pit of pride, chronic illness... and so many others!

One day, the prophet Jonah (whose story is told in the Bible, in the book that bears his name) found himself in a place so dark and so deep because of his actions that all he could do was talk to God and pray. He was trapped in the belly of a great fish that had swallowed him, on the high seas, in the depths of the waters... And there he learned the lesson of obedience. It was there that he had the most transforming encounter with God of his life! He had been sucked into that deep pit like a stubborn man. But three days later, he was thrown on the beach like a broken man, ready to go wherever God sent him.

Dear reader, my prayer is that you will allow the Lord to find and reach you wherever you are. Believe that He sees you, that He cares for you, that He loves you, and that He has His right hand outstretched to draw you closer to His heart.

Don't stay in the pit! Accept the help you need to get out! Maybe God will give you a sympathetic friend to help you, or 30 people to intercede for you before the throne of grace, pulling hard on the ropes so you can get free!

Alone with God

Prayer

Lord, I thank You for Your infinite love. Thank You for caring for me even though You know all about me, who I am, where I am, and why I'm here.

Today, by faith, I look up at the exit of this pit I have fallen into, and I realize that You are the only one who can bring me out. I hunger and thirst for you, O God.

I'm ready to take the opportunity that You have given me and let You pull me out of this discouragement, this darkness, this disappointment...

I know that the blood of Jesus Christ has the power to cleanse me from all sin. By faith, I receive Your forgiveness, and by faith I now begin to climb out of the pit...

In the name of the Lord Jesus Christ.

Part V

Prophets

Chapter 1
Some nameless prophets

The ministry of priests

The messages from God that the Old Testament prophets gave were generally addressed to the people as a whole, but also, in many cases – such as those we are about to consider – to individuals.

It was a tough ministry. The lives of these men of valor were often at risk. In various situations, they represented the message with their own bodies, in addition to the words they spoke, as in the example of the prophet who asked to be punched. In this way, he became a kind of "visual aid" to support the message that the Lord wanted to convey to the king.

Many prophets have lost their lives for being God's voice in the world. The Lord Jesus Christ referred to this fact in his lament over Jerusalem, which will be quoted below.

In the midst of the treasures that we discover as we go through God's Word, we find little shiny stones that, upon examination, reveal themselves to be of great value. This is the case of the many extraordinary men whose names are not revealed to us in the Scriptures.

In this chapter we will reflect very briefly on the life and mission of some prophets mentioned in the Bible without any reference to their name. They are part of the great multitude of anonymous people who have followed and served the Lord through the ages, over the millennia, while we await *"the blessed hope and the appearing of the glory of our great God and Savior, Christ Jesus"* (Titus 2:13).

In the following diagram, some information about each of the five prophets we will mention here is written down, so that we can identify them more easily:

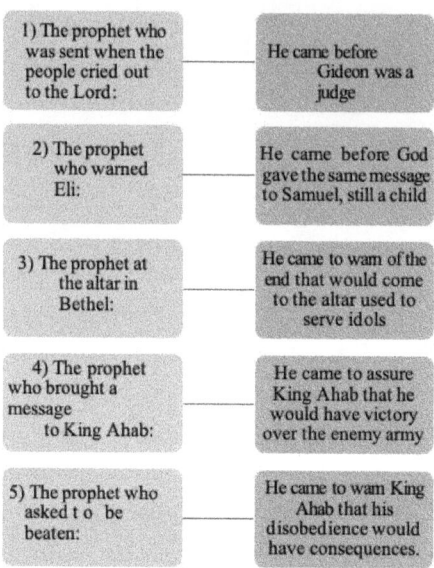

1. The prophet sent when the people cried out to the Lord for deliverance

BIBLICAL NARRATIVE

Now it came about when the sons of Israel cried to the LORD on account of Midian, that the LORD sent a prophet to the sons of Israel, and he said to them, "Thus says the LORD, the God of Israel, 'It was I who brought you up from Egypt and brought you out from the house of slavery. I delivered you from the hands of the Egyptians and from the hands of all your oppressors, and dispossessed them before you and gave you their land, and I said to you, "I am the LORD your God; you shall not fear the gods of the Amorites in whose land you live. But you have not obeyed Me.'" (Judges 6:7–10)

After the great victory of the army of Israel over the Canaanites, led by the judge Deborah and the general Barak, the people quickly forgot the Lord's deliverance and faithfulness. Only 40 years had passed – a short time in the life of a nation: 40 years of peace. But just as *"A sow, after washing, returns to wallowing in the mire"* (2 Peter 2:22), the nation fell back into idolatry. Again, the Lord got their attention by allowing another people, the Midianites, to invade, rob, and oppress Israel until their cry for help moved God's heart.

In His omniscience, the Lord had already chosen a man to deliver the nation from this yoke. But before He recruited that man, He sent a prophet with a harsh message. It was as if the people had been *"weighed on the scales and found deficient"* (Daniel 5:27)

Basically, the message of the anonymous prophet had three important points:

1.   to explain to the Hebrews why they were under the oppression of a nomadic, hungry, and cruel people;
2.   to remind them who the Lord was, their God – the Deliverer, the Almighty;
3.   to remind them of the command they had received from Him not to fear, much less worship, the idols of the land they were occupying.

The people understood the message perfectly and continued to cry out for the Lord's intervention until He called Gideon and his army of 300 hand–picked men, equipped with trumpets, pitchers, and lighted torches, to deliver Israel from the yoke of the Midianites.

The nameless prophet was the voice of God speaking in sorrow to His people.

The centuries passed, and one day in Jerusalem, the Lord Jesus Christ expressed His sorrow for the nation in these words:

"O Jerusalem, Jerusalem, the city that kills the prophets and stones those sent to her! How often I wanted to gather your children together, just as a hen gathers her brood under her wings, and you would not have

it!" (Luke 13:34)

Then He prophesied what would happen to His people:

"Behold, your house is left to you desolate; and I say to you, you will not see Me until the time comes when you say, 'BLESSED IS HE WHO COMES IN THE NAME OF THE LORD!'" (Luke 13:35)

The Lord never fails to hear us when we bring our petitions to Him. But how many times have we cried out to Him to come and deliver us in a supernatural way when we have placed ourselves under the yoke that oppresses us! And how have we done this? By rejecting what we knew to be the right way, God's will, and choosing our own way, in our own way, in our own time, according to our own will!

"God be gracious to us and bless us,
And cause His face to shine upon us— Selah" (Psalms 67:1)

2. The prophet who warned Eli

BIBLICAL NARRATIVE

Then a man of God came to Eli and said to him, "Thus says the LORD, 'Did I not indeed reveal Myself to the house of your father when they were in Egypt in bondage to Pharaoh's house? Did I not choose them from all the tribes of Israel to be My priests, to go up to My altar, to burn incense, to carry an ephod before Me; and did I not give to the house of your father all the fire offerings of the sons of Israel? Why do you kick at My sacrifice and at My offering which I have commanded in My dwelling, and honor your sons above Me, by making yourselves fat with the choicest of every offering of My people Israel?' Therefore the LORD God of Israel declares, 'I did indeed say that your house and the house of your father should walk before Me forever'; but now the LORD declares, 'Far be it from Me—for those who honor Me I will honor, and those who despise Me will be lightly esteemed. Behold, the days are coming when I will break your strength and the strength of your father's house so that there will not be an old man in your house. You will see the distress of My dwelling, in spite of all the good that I do for Israel; and an old man will not be in your house forever. Yet I will not cut off every man of yours from My altar so that your eyes will fail from weeping and your soul grieve, and all the increase of your house will die in the prime of life. This will be the sign to you which will come concerning your two sons, Hophni and Phinehas: on the same day both of them will die. But I will raise up for Myself a faithful priest who will do according to what is in My heart and in My soul; and I will build him an enduring house, and he will walk before My anointed always. (1 Samuel 2:27–35)

This prophet came to warn the priest Eli of the consequences that his permissiveness as a father and the Lord's priest would have on his lineage.

- When he knew the sins of his sons regarding the sacrifices offered on the altar of burnt offering, he did not rebuke them;
- Knowing how disrespectfully they handled the offerings, he didn't correct them;
- Knowing their greed and dishonesty, he allowed himself to be overcome by sloth and allowed the name of the Lord to be dishonored.

That's why the Lord decided to:

- remove Eli's descendants from the ministry;
- not allow any of them to live long: they would all die *"in the prime of life"*;
- raise up another man to replace him.

In the above quoted passage from the Word of God, we find a basic principle for our daily and ministerial conduct: "those who honor Me I will honor". Reader, cherish this principle like a jewel. Use it as a cornerstone in your life. Use it as a parameter for self–evaluation throughout your faith and professional career. Make it a sieve for your attitudes, words, and decisions. Believe that it is a seed that will bear much fruit and yield a profit higher than you can imagine!

3. The prophet of the altar of Bethel

BIBLICAL NARRATIVE

Now behold, there came a man of God from Judah to Bethel by the word of the LORD, while Jeroboam was standing by the altar to burn incense. He cried against the altar by the word of the LORD, and said, "O altar, altar, thus says the LORD, 'Behold, a son shall be born to the house of David, Josiah by name; and on you he shall sacrifice the priests of the high places who burn incense on you, and human bones shall be burned on you.'" Then he gave a sign the same day, saying, "This is the sign which the LORD has spoken, 'Behold, the altar shall be split apart and the ashes which are on it shall be poured out.'" Now when the king heard the saying of the man of God, which he cried against the altar in Bethel, Jeroboam stretched out his hand from the altar, saying, "Seize him." But his hand which he stretched out against him dried up, so that he could not draw it back to himself. The altar also was split apart and the ashes were poured out from the altar, according to the sign which the man of God had given by the word of the LORD. The king said to the man of God, "Please entreat the LORD your God, and pray for me, that my hand may be restored to me." So the man of God entreated the LORD, and the king's hand was restored to him, and it became as it was before. (1 Kings 13:1–6)

This unnamed prophet traveled from Judah to Bethel with a stern message to deliver. This message was about idol worship at an altar in Bethel, where only the Lord should be worshipped – because only the Lord is God!

He entered the scene with a powerful testimony to prove to King Jeroboam that his words were from the Lord:

"This is the sign which the LORD has spoken, 'Behold, the altar shall be split apart and the ashes which are on it shall be poured out.'" (1 Kings 13:3)

His message was very specific: someone from the royal house of David, whose name would be Josiah, would be born to put an end to the idolatry being practiced on that altar! King Jeroboam responded by reaching out his hand to order the prophet's arrest and possible execution. But the Lord intervened on behalf of His faithful servant:

- The king's hand *"withered away"* – he remained with his arm outstretched, as if he were a stone statue;
- The altar broke before his eyes;
- The ashes spilled onto the ground.

The Lord had spoken. There was the proof: the broken altar, just as the man of God had just prophesied!

The king responded again: *"Please entreat the LORD your God, and pray for me, that my hand may be restored to me."*

The Lord, the Almighty God, was not the God of King Jeroboam, but He was the God of the prophet. If He had the power to split an altar, He must also have the power to heal his arm, Jeroboam thought. And indeed, the power of the Lord was once again manifested! The prophet's intercession was answered and the king's arm was restored. He invited the prophet to come to his house and eat with him, but the man of God had no such orders. So he declined the invitation and went on his way.

Just as the Lord Jesus Christ proved that He was God's Messiah – the power to perform miracles, forgive sins, heal, resurrect, multiply bread and fish, walk on the water of the sea, calm the wind and waves, resist all the temptations of Satan, escape through the crowd that wanted to kill Him, etc. – this prophet also presented evidence that He had been sent by God.

Reader, when the Lord sends us somewhere or to someone, it is important that we are sure that we can present the credentials of a clean and consecrated life. We should not preach what we don't practice or talk about realities that are not part of our daily experience of walking with the Lord.

One way or another, the people who hear us will end up knowing whether we have been with the Lord or not, just as happened to the apostles Peter and John (Acts 4:13) many centuries ago!

4. The prophet who brought a message to King Ahab

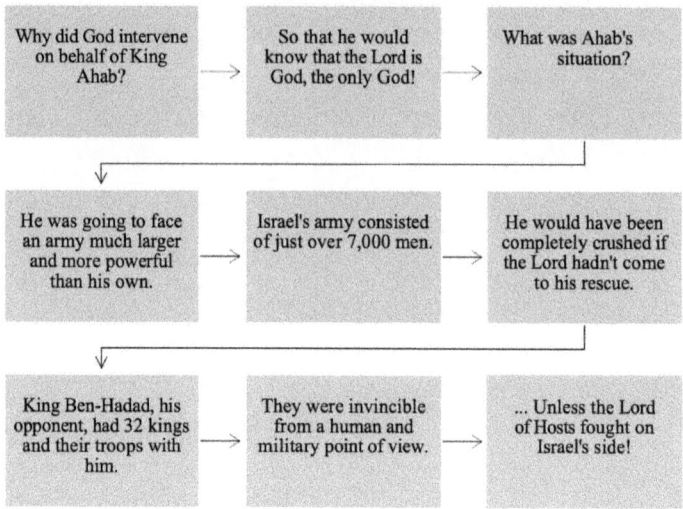

Why did God intervene on behalf of King Ahab? →	So that he would know that the Lord is God, the only God! →	What was Ahab's situation?
He was going to face an army much larger and more powerful than his own. →	Israel's army consisted of just over 7,000 men. →	He would have been completely crushed if the Lord hadn't come to his rescue.
King Ben-Hadad, his opponent, had 32 kings and their troops with him. →	They were invincible from a human and military point of view. →	... Unless the Lord of Hosts fought on Israel's side!

BIBLICAL NARRATIVE

Now behold, a prophet approached Ahab king of Israel and said, "Thus says the LORD, 'Have you seen all this great multitude? Behold, I will deliver them into your hand today, and you shall know that I am the LORD.'" Ahab said, "By whom?" So he said, "Thus says the LORD, 'By the young men of the rulers of the provinces.'" Then he said, "Who shall begin the battle?" And he answered, "You." (1 Kings 20:13–14)

Ahab was a wicked king, one of the worst to ever rule Israel, yet the Lord was very merciful to him.

On a day of need, before the start of a battle, an anonymous prophet brought the king a word of encouragement from God: he would win the battle, but he had to take the initiative to start it. And so it was.

The Lord is willing to go to great lengths and spares no effort to:
save a soul;
• reveal himself to someone;

- rescue a wounded person from the battlefield – particularly those hit by "friendly fire";
- give someone another chance.

He did it for Ahab. He continues to do it for millions of people around the world. You are no exception! See if there is a special message from the Lord for you in the pages of Scripture, or even in this book.

5. The prophet who asked to be beaten

BIBLICAL NARRATIVE

Now a certain man of the sons of the prophets said to another by the word of the LORD, "Please strike me." But the man refused to strike him. Then he said to him, "Because you have not listened to the voice of the LORD, behold, as soon as you have departed from me, a lion will kill you." And as soon as he had departed from him a lion found him and killed him. Then he found another man and said, "Please strike me." And the man struck him, wounding him. So the prophet departed and waited for the king by the way, and disguised himself with a bandage over his eyes. As the king passed by, he cried to the king and said, "Your servant went out into the midst of the battle; and behold, a man turned aside and brought a man to me and said, 'Guard this man; if for any reason he is missing, then your life shall be for his life, or else you shall pay a talent of silver.' While your servant was busy here and there, he was gone." And the king of Israel said to him, "So shall your judgment be; you yourself have decided it." Then he hastily took the bandage away from his eyes, and the king of Israel recognized him that he was of the prophets. He said to him, "Thus says the LORD, 'Because you have let go out of your hand the man whom I had devoted to destruction, therefore your life shall go for his life, and your people for his people.'" So the king of Israel went to his house sullen and vexed, and came to Samaria. (1 Kings 20:35–43)

This prophet's message came wrapped up in a performance, as if it were a play.

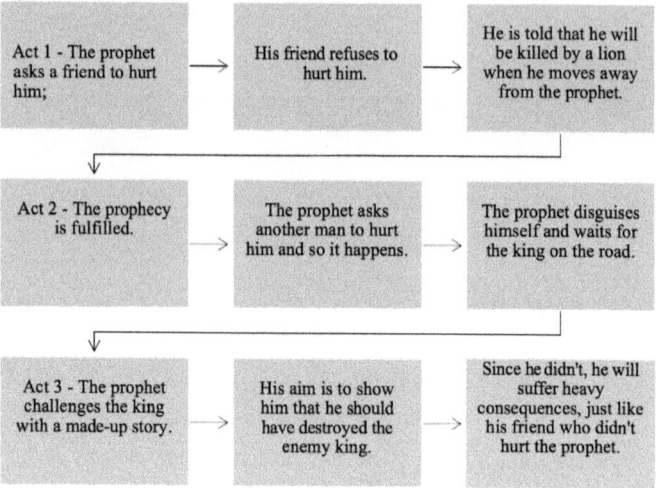

The Lord's rebuke of King Ahab was for his failure to carry out the mission he had been given: to destroy the enemy king. Instead, Ahab befriended him, even though he was aware of all the damage he had already done to Israel.

Sometimes we dare to think that we are more righteous than God Himself! At that time, Israel was still taking possession of the land the Lord had promised them. There were kings and nations that had to be driven out of their territory. Why? Because God could no longer tolerate the sin of those nations!

Many of the nations that were fought against and driven out of the Promised Land were cruel peoples who sacrificed children on the altars of their idols, who gave home to their practices of witchcraft, incest, and brutality. It is not true that the Lord simply carried out "ethnic cleansing". He gave all these nations time – 400 years – to change their ways. Israel was placed in their midst to serve as a testimony to them and to show them what happens – the blessings that fall on individuals – when a nation trusts in the living and true God!

The Lord was also setting the stage for the coming of the Messiah – the Savior of all peoples, nations, and tongues, according to the promise made to Abraham that in Him all the families of the earth would be blessed.

"And I will bless those who bless you,
And the one who curses you I will curse.
And in you all the families of the earth will be blessed." (Genesis 12:3)

"Your descendants will also be like the dust of the earth, and you will spread out to the west and to the east and to the north and to the south; and in you and in your descendants shall all the families of the earth be blessed." (Genesis 28:14)

"The Scripture, foreseeing that God would justify the Gentiles by faith, preached the gospel beforehand to Abraham, saying, "ALL THE NATIONS WILL BE BLESSED IN YOU." (Galatians 3:8)

God, "who desires all men to be saved and to come to the knowledge of the truth." (1 Timothy 2:4)

A moment alone with God

Almighty God,

Help me to discern Your voice from all the other voices I hear in this world. Let me listen to You so that I may have something to say to those who listen to me.

Help me to read Your Word and discover Your will for my life.

I ask You to transform me into a faithful, obedient and diligent servant, ready to take Your message of salvation to whomever You send me.

In this moment, I look away from my own weaknesses and look only to Your power.

I'm counting on you, Lord. I want to go so far as to tell You that You can count on me, too.

In the name of the Lord Jesus Christ, I pray.

Chapter 2
A prophet with a mission

- Historical data -

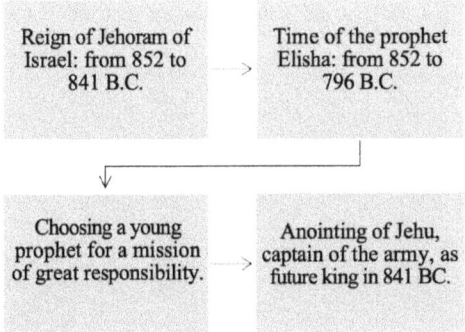

Biblical narrative

Now Elisha the prophet called one of the sons of the prophets and said to him, "Gird up your loins, and take this flask of oil in your hand and go to Ramoth–gilead. When you arrive there, search out Jehu the son of Jehoshaphat the son of Nimshi, and go in and bid him arise from among his brothers, and bring him to an inner room. Then take the flask of oil and pour it on his head and say, 'Thus says the LORD, "I have anointed you king over Israel."' Then open the door and flee and do not wait."

So the young man, the servant of the prophet, went to Ramoth–gilead. When he came, behold, the captains of the army were sitting, and he said, "I have a word for you, O captain." And Jehu said, "For which one of us?" And he said, "For you, O captain." He arose and went into the house, and he poured the oil on his head and said to him, "Thus says the LORD, the God of Israel, 'I have anointed you king over the people of the LORD, even over Israel. You shall strike the house of Ahab your master, that I may avenge the blood of My servants the prophets, and the blood of all the servants of the LORD, at the hand of Jezebel. For the whole house of Ahab shall perish, and I will cut off from Ahab every male person both bond and free in Israel. I will make the house of Ahab like the house of Jeroboam the son of Nebat, and like the house of Baasha the son of Ahijah. The dogs shall eat Jezebel in the territory of Jezreel, and none shall bury her.'" Then he opened the door and fled. (2 Kings 9:1–10)

A disciple of the prophets

"(...) Elisha the prophet called one of the sons of the prophets (...) So the young man, the servant of the prophet (...)"

If you're as inquisitive as I am, reading this passage (2 Kings 9:1–10) will raise some questions about the man who was sent to anoint Jehu, a captain in Israel's army, as the nation's future king. This young man will be the subject of our study in this chapter. We will try to find answers to some questions:

- *Who was this young man?* His name is unknown, as are the names of his ancestors. To us he is anonymous, to God he is a brave servant.
- *Why was he in a school of prophets?* I believe that the only people who go to a school of prophets are those who want to know the Lord, His Word, and how to serve Him better.
- *Why was he chosen for the mission proposed by the prophet Elisha?* The choice couldn't have been random. It never is. We may not

even fully understand why we invite a particular person for a short mission or a longer ministry, but nothing is random. The prophet Elisha would have sensed the Lord's direction to choose this young man and not another, "for man looks at the outward appearance, but the LORD looks at the heart." (1 Samuel 16:7)

- *What qualities distinguished him from the other disciples of the prophets?* We know only that he was a young man with the attitude of an apprentice, trustworthy, courageous, and obedient, which are invaluable qualities in youth. The Scriptures don't record any objection he might have had to the mission the prophet Elisha proposed to him. There was no "what if?": "What if the captain won't receive me? What if the other captains question me? What if I can't remember everything I'm supposed to say? What if I can't run fast enough?" Nothing! The young man took heed of what he was told, accepted the mission, and set out to do it!

Get ready

"Gird up your loins..." – This was Elisha's first command to the young prophet. This command, reader, in a figurative sense, can be given to us at any time in our lives. It means something like:

- There's a job for you! Dress for your mission – nothing fancy, but with care, cleanliness and no sloppiness;
- Get rid of burdens and obstacles: *"Therefore, since we have so great a cloud of witnesses surrounding us, let us also lay aside every encumbrance and the sin which so easily entangles us, and let us run with endurance the race that is set before us, fixing our eyes on Jesus (...)"* (Hebrews 12:1–2).
- Get ready to walk and to run! Living by faith is a long race. It won't be done without perseverance, endurance and resilience;
- *"(...) if anyone competes as an athlete, he does not win the prize unless he competes according to the rules"* (2 Timothy 2:5), said the apostle Paul to his beloved and faithful young disciple, Timothy. These

are words that make us reflect on how we serve the Lord and how we run the race of faith.

The expression "to gird up the loins" referred to preparation for physical labor without the limitations of the tunic. A belt, girdle, or even a strip of cloth was worn over the tunic to hold it in place and shorten its length so that it was possible to walk, run, or work for long periods of time without the hindrance of a long tunic.

As for the distance to be covered, we know that there was a school of prophets in Bethel and another in Jericho, and they certainly weren't the only ones. The Bible doesn't tell us where the young prophet set out from to reach Ramoth–gilead, but it is likely that he walked many miles. He did so with his *"loins girded,"* also because he had to flee as soon as his mission was accomplished.

The young man did not fear that he might not be sufficiently prepared for the responsibility of going to anoint Captain Jehu as the future king of Israel. Even if this doubt had crossed his mind, we don't read about it – he didn't express it. He accepted Elisha's orders and prepared to carry them out like a good soldier before his commander!

Was he chosen among all the other disciples of the prophets because he was the bravest? The most agile and the fastest in races? The most holy? The wisest? The strongest? The one who answered first?

The one who couldn't go unnoticed? Or, on the other hand, was he the quiet, introverted, attentive, meditative one, with an inner strength generated by a solid faith in God? Would he be the disciple who needed the most encouragement? What do you think about these questions? Was the mission the prophet Elisha gave this young man a kind of reward for his merits and efforts? Or was it a stepping stone to greater and higher flights?

The value of praise and encouragement

If we're honest, we can admit that we all need to hear a word of appreciation and recognition for our work or our character from time to time. The people who work with us on the same project – family, church, missionary organization, business, and many others – also need a compliment from time to time. Don't be afraid to give sincere praise. What the other person does with the praise they receive is not your concern. It's between them and God. In Proverbs 27:21 we find these words of wisdom: "As the fining pot for silver, and the furnace for gold; so is a man to his praise." (KJV). Praise should produce in us a greater sense of responsibility. It's not a crown to exalt us, it's a little oil in the engine to make it run better!

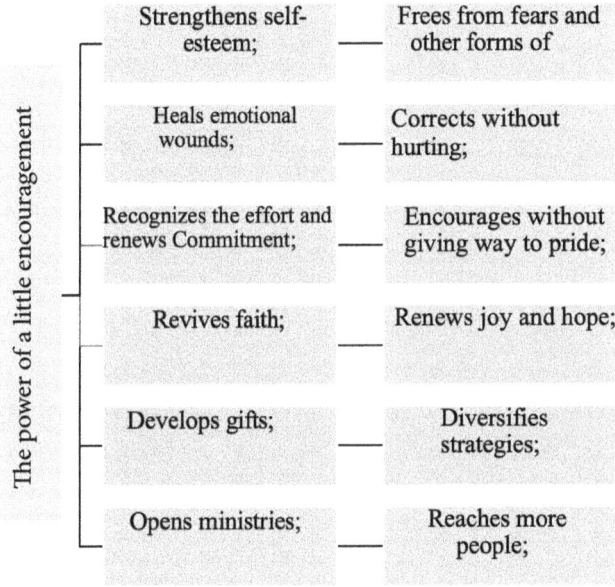

Olive oil

"(...) take this flask of oil in your hand and go (...)"

- The oil would be poured over Captain Jehu's head, while the prophet–disciple conveyed to him the word of God: *"Thus says the LORD, the God of Israel, 'I have anointed you king over the people of the LORD, even over Israel'"* (2 Kings 9:6).
- As a symbol of the Holy Spirit, we can draw from Elisha's words

in sending the young prophet the implication that no one should be commissioned to serve the Lord without bringing with them a *"flask of oil"* – the anointing of God's Holy Spirit.

- Without the oil, there would be no anointing of the new king. In the same way, without the anointing of the Holy Spirit on the ministry – whatever it may be – there will always be something missing, that special touch of the Lord's presence, the anointing.

- No service, however simple it may seem, can do without the anointing of the Holy Spirit! It is He who makes all the difference in the quality of the ministry,** in the commitment and consecration of those who carry it out, in the scope of the work, in the fruit that will come afterwards.

Let us make sure that our inner man is not only *"sealed in Him with the Holy Spirit of promise"* (Ephesians 1:13), but also filled with the joy and power of the Holy Spirit, anointed by the Holy Spirit, set apart and consecrated by the Holy Spirit for the service and worship of the Lord! What a difference it will make in our lives and what a blessing it will be in the lives of others when we live like this!

Back to our young prophet: with the jar of oil in his hand, he really had to go! He could no longer sit quietly at the feet of the teachers, absorbing every word like a sponge, keeping all the teachings in his mind and heart like someone guarding a treasure. That position was no longer possible – he had the oil in his hand and had to carry out the task that had been given to him! He had received the call of his life – he would forever be the young prophet who would anoint a new king! He had been honored with an extraordinary mission – one that

---

** By *ministry* we mean the exercise of all our talents and gifts, wherever we go, at all hours of the day and night, and in all circumstances. It encompasses our service in the Church, in the family and in society, without any duality of *modus operandi*. It's our way of being, living, and acting wherever we are, whether in the light of a spotlight or in the darkness of a corner of our home, alone in the car, in the middle of traffic, alone in the office, alone at a meal, or alone at a task that no one sees but the Lord.

awaited him and depended on him, with no substitute. The future of the nation was in his hands, in that vessel of oil. It was time to go!

A word to the reader

When we understand what the Lord's call is for our lives, it's time to go! It's time to "gird up our loins," pick up the "jar of oil," and be willing to pay the price to serve a higher purpose: God's plan for us!

We will hardly be called to anoint a king, or even to deliver a secret message of national significance, as was the case with the young prophet: the Lord used the king he anointed to bring judgment on the house of Ahab, to cleanse the nation of his evil influence, to avenge the destruction of the Lord's prophets at Jezebel's command, and finally to extinguish the cult of Baal. There was an Israel before the young prophet's mission and an Israel after, such were the consequences of his act of obedience on the country!

We, who are also God's anonymous ones, not even prophets, will always be called to various tasks and missions that are within our reach and that will make a difference to many people. Here are some examples:

- Encouraging someone who is down or unmotivated;
- Praising a job well done;
- Recognizing someone's qualities and effort, with words and attitudes;
- Giving someone who has already failed a few times another chance – isn't that what the Lord does with us?
- Forgiving someone who has hurt us, harmed us or destroyed a dream of ours;
- Sharing faith in the Lord Jesus Christ with those who don't know Him as Savior;
- Sharing the bread we have, the water, the house, the car, the knowledge and other blessings that the Lord has given us;
- Building the church with our gifts and talents, under pastoral authority and, above all, at the Lord's command;

- Being the right arm of a tired worker;
- Praying in a disciplined and constant way;
- Praying faithfully for anyone who sends us an "S.O.S. prayer";
- Fighting in prayer for someone's life;
- Taking the time to help someone;
- Being present in moments of joy and in moments of pain in the lives of our "neighbors";
- Being a facilitator;
- Being a peacemaker, a kind of holy fireman of God, putting out fires, even at the risk of our own life and reputation;
- Being active, pulling the plow instead of being a dead weight that slows down the progress of the whole group;
- Standing firm in the storms of life: not abandoning the boat, but continuing to row alongside the other faithful rowers;
- Accepting God's will and making it a pulpit, a bridge to reach other people, or whatever the Lord shows us that should be done;
- Always learning and never ceasing to be teachable;
- Developing the gifts and talents that God has entrusted to us;
- And all the other specific missions that the Lord puts in front of us.

A moment alone with God

Prayer

Lord, I come to worship You for being my God and my Savior! Thank You for the privilege of serving You.

Thank You for every assignment and every mission.

Thank You for every opportunity and every responsibility You have given me.

I ask You to empower me to do more, better and faster, giving You all the glory!

I ask You to help me be ready to "gird my loins", "take up the jar of oil" and go where You want me to go.

I ask You to make me a useful servant and an encourager. In the name of the Lord Jesus Christ.

Part VI

Messenger men

Chapter 1
Job's 4 messengers

If you are like most people, when you read the first few chapters of Job's story in Scripture, in the book that bears his name, your attention will be focused on the following aspects:

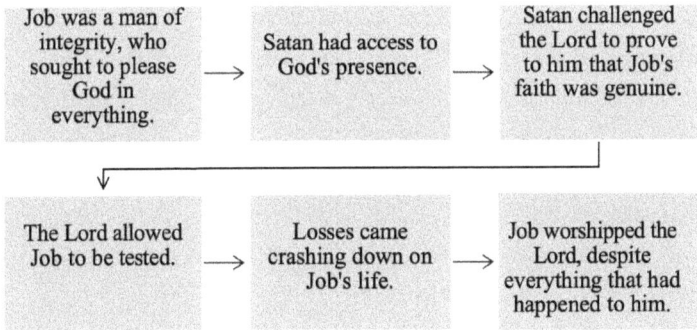

It's only natural that a wave of indignation should hit us when we see Satan's audacity! After all, he shouldn't even be allowed near God's holy angels, let alone the Lord Himself! What's more, he dared to defy the Lord in order to prove Job's faith. He was "sure" that Job would fail the test, and that would give him great pleasure! He was "sure" that Job was a man who served the Lord because of the blessings he had received from the Most High. He thought that there was a direct proportion between the blessings received and the worship given, so he made his cruel suggestion, with hints of a bet:

- He bet God that Job would fail completely;
- he bet God that Job, who was "blameless, upright, fearing God and turning away from evil", would curse the Lord the moment he lost his possessions and his family;
- he bet God that, if he also lost his health, Job would fall into the deep pit of rebellion and would never be who he was again.

Well, we've read the story and we know how it ends. Knowing the goodness and mercy of the Lord, even before we read the happy ending of Job's life, we had to know that God's triumph would be total and resounding! But Satan didn't know it. He stood by in anticipation after each of Job's blows. He doesn't understand God's love. He refuses to accept that the Lord can do anything, that He has the last word, and that his victory is guaranteed long before the battle begins! Nor does he believe in man's true, unselfish love for the Lord. He would be defeated, but he wouldn't accept that thought. He would be utterly embarrassed and crushed by the testimony of this man of integrity that he couldn't bear to watch! He would have the opportunity to see Job's tears, of course, but he would also have the opportunity to hear him utter the phrase that was produced by his unbreakable faith: "The Lord gave and the Lord has taken away. Blessed be the name of the Lord." Job worshipped the Lord at the end of all the bad news! Job remained upright, righteous, God–fearing and turned away from evil!

Reader, this is very difficult! I'm sure you've found yourself in situations where you had a lot of unanswered questions, or at least ones that weren't easy to understand. How I understand you!

It's also possible that the reader has already felt so hurt that he hasn't even tried to question anything... I understand that too. However, there are some simple things we can always do:

- *Talk to God about the problem. Talk to Him much more than to people who can do little or nothing for you; if you have someone who can understand, support, encourage, and pray for you – that person is a precious treasure! If you have a relationship of this caliber, hold*

*on to it! But don't share your pain with those who are just waiting
for an opportunity to unroll their own personal list of regrets and
burden you even more!*

- *Ask the Lord to help you get through the test. He knows your
  structure, He knows that you are just dust (Psalm 103:14), and in
  His faithfulness He will not test you beyond your ability to bear (1
  Corinthians 10:13);*

- *Don't be angry, because anger is a gateway to other negative feelings
  and to opportunistic unclean spirits;*

- *Do not be angry with God, for if He allows a wound, He will also
  bind it! (Job 5:18; Hosea 6:1)*

- *Remembering the blessings the Lord has already poured out on your
  life is a good strategy for reviving your faith and renewing your hope.
  He who has been faithful in the past remains faithful now and will
  be faithful to the end!*

- *Think about why the Lord allowed a particular tribulation to touch
  your life. Ask "what for" more often than "why," though your "whys"
  can't hurt God's heart.*

- *Read the Word. Don't sulk with God! If you stop reading His love
  letter to humanity every day, it's your loss, not the Lord's!*

- *Don't give God ultimatums. Time is in His hands, not yours! Absolute
  power, complete wisdom and the ability to help, to solve problems,
  to see the end of things before they even begin, belongs to the Lord!*

- *Rest in God's promises. Affirm and reaffirm: "Come what may, I am
  Yours and You will be my Lord until death and forever and ever!"*

- *Follow Job's attitude: "Though he slay me, I will hope in him." (Job
  13:15)*

The 4 messengers

*[A] messenger came to Job and said, "The oxen were plowing and the
donkeys feeding beside them, and the Sabeans attacked and took them.
They also slew the servants with the edge of the sword, and I alone*

have escaped to tell you." While he was still speaking, another also came and said, "The fire of God fell from heaven and burned up the sheep and the servants and consumed them, and I alone have escaped to tell you." While he was still speaking, another also came and said, "The Chaldeans formed three bands and made a raid on the camels and took them and slew the servants with the edge of the sword, and I alone have escaped to tell you." While he was still speaking, another also came and said, "Your sons and your daughters were eating and drinking wine in their oldest brother's house, and behold, a great wind came from across the wilderness and struck the four corners of the house, and it fell on the young people and they died, and I alone have escaped to tell you."

Then Job arose and tore his robe and shaved his head, and he fell to the ground and worshiped. He said,

"Naked I came from my mother's womb,
And naked I shall return there.
The LORD gave and the LORD has taken away.
Blessed be the name of the LORD."

Through all this Job did not sin nor did he blame God. (Job 1:14–22)

- The first messenger brought the terrible news that Job had lost all his oxen and donkeys, as well as the servants who tended them, in a sudden attack by the Sabeans;
- The second messenger came to say that he had seen fire fall from the sky, killing the sheep and the shepherds who guarded them;
- The third messenger reported that Job's camels had been taken by three bands of Chaldeans, who not only stole all the camels but also killed with their swords the servants who were tending the animals, just as the Sabeans had done;
- The fourth messenger came with the worst news of all: a great wind from the desert had blown down the house of Job's eldest son, where all his other sons were, ten in all!

Job's losses reached land daughters the very core of his soul. Soon, without a messenger, an extremely serious illness would come and weaken his body terribly. Then Job would lose what he had left: the support of his wife, who had lost all her possessions, just like him, all her children, just like him, but couldn't bear to look at her husband's physical suffering. She had given up hope for better days...

The losses suffered by Job and his messengers

Job was a man of integrity, an understanding employer who treated all his servants with kindness and generosity, who didn't shortchange any of them in their wages, but rather made sure that everyone had what they needed. This is the way of every man of integrity who has people working in their business or on their farm, ranch, or other property.

Everything he had lost was partially recoverable, perhaps through many years of work and good business, but the lives lost were lives lost. The disaster that struck Job's home also affected the homes of his servants: entire families were left without a father or husband, without a son or brother, without food or means of survival. In other circumstances, Job would have helped this anonymous crowd of his servants' relatives, but at this point he could do nothing for them. This fact also weighed heavily on his heart.

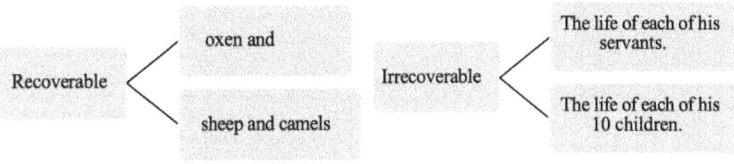

The four messengers who brought Job the bad news had something in common: a terrifying experience.

- They had witnessed the attacks of people and gangs who stole all the animals;
- They had witnessed the violent deaths of their companions;

- They saw the terrifying fire that fell from the sky and destroyed everything, both men and flocks;
- They heard and saw the impact of the great desert wind that reduced the eldest son's house to rubble. The messenger who brought this news had been in the house, where a lively feast was going on, in an atmosphere of brotherly love and joy! Now the ten children were dead.

There was a common denominator in the lives of these men – they had been spared, they had survived to bring the news to Job. They almost felt guilty for surviving! Why them and not any of the others?

This is how Satan works: he uses people or even elements of nature to carry out his cruel and insidious plans! It's no accident that the Lord Jesus Christ said, "The thief comes only to steal and kill and destroy" (John 10:10).

God's choice

Considering that the Lord is the Almighty God and that everything He allows to happen has a very well-defined purpose, we believe that the choice of the 4 survivors was not random. They remain anonymous to us, but they were never anonymous to God! They certainly had qualities that the Lord wanted in each of the messengers so that the facts could be reported accurately. They were honest, attentive, and good observers. In none of the cases do we read that the man who escaped disaster was dominated by inner feelings of fear or self-pity. They were focused on Job, their master, his losses and his pain. Each of them concludes the delivery of bad news by saying: "I alone have escaped to bring you news." It was a mixed feeling of responsibility and grief. You can almost hear them thinking aloud: "It would have been better to die than to have to bring this news to Job!" But God always makes the right decision!

The reading of Job chapter 1, verses 13 through 19, indicates that the four messengers arrived one after the other, without enough time for Job to pull himself together and prepare for the next piece of news. It seems that the four tragic events happened almost simultaneously! Satan waited. He waited as a predator waits for the prey to devour. He waited for the day when Job's ten children would be together in the same house celebrating life, he waited for the oxen to be in the field and the donkeys too, he waited for the sheep to be in the pasture, and he waited for the strong desert wind to blow and sweep away everything that belonged to this man of integrity.

Reader, your adversary – and mine too – acts like this: he waits for the right conditions to arise to attack us where it hurts the most! And when he sees us writhing in pain, he doesn't stop, he continues to attack with his cruelty, his mercilessness, and his unfaithfulness! Don't think he's going to feel sorry for you someday! He won't! He is essentially evil, he plays dirty, he deceives, and he will not give up until he is convinced that he has won, or until the Lord stretches out His hand and says, "Enough"! Pray that the Lord will not accept any proposal from Satan against you or your family! Remember that the Lord Jesus Christ is your faithful Advocate at the right hand of the Father. Remember that the Holy Spirit perfects your prayers and intercedes for you before God the Father. But the Holy Spirit will not perfect a prayer that hasn't been said! Remember this.

Walk with the Lord, live for Him, serve Him, love Him, and worship Him! Stay close to Him. And when the Lord puts you to the test to make you a better man, to cut that rare diamond in you, see if you come out approved, to the glory of God!

Pray a lot! Pray always and everywhere! Pray in all circumstances! Pray and don't give up! Pray even when you think you don't know how to pray well. Don't be deceived! Pray and the Lord will hear you, answer you, deliver you and bless you, in the right measure and at the right time!

And if the Lord chooses you to be a survivor of a disaster, or to bring news to someone, perhaps even in the form of a sermon from the pulpit, do it faithfully! There is plenty of good news that should never be kept to ourselves! There is news of salvation and hope for a world tired of bad news. There are people to whom only you have privileged access. Be bold and bless the lives of everyone you can with the good news of God's love and forgiveness!

A moment alone with God

Prayer:

Lord,

I believe that everything that happens to me works together for my good, because You are my God, whom I love with all my heart and with all my strength.

I believe that the Holy Spirit perfects my weak prayers, so I ask You to help me pray always!

I believe that the Lord Jesus Christ is my Advocate and that He intercedes for me so that I may have peace.

I believe that You have called me and saved me for a good, high and holy purpose. Here I stand again before you, and here I lift up my supplication:

- Empower me to be the person You want me to be;
- Help me to be faithful in communicating the good news to others and wise to encourage the discouraged;
- Give me boldness in speaking, holiness in witnessing my life and the opportunity to help anyone You put in my path.

In the name of the Lord Jesus Christ.

Moment of reflection

I will not lower my arms,
nor slow down my walk with Christ!
I won't stop praying, no matter what,
Whether I'm tested or whatever comes.
If Satan tries to undermine my faith,
To stop me from walking or standing,

I will run to you, Lord, to your altar,
As any river flows to the sea.

In your holy presence I can rest,
And tell you all my anguish and pain,
Certain that Your immense and strong love
Will come for me, will come to heal me,
And will carry me in sweet safety
Beyond the waves, where there is calm,
After the storm and the violent force
Of the many struggles, of the many waters.

When the night has passed and the sorrows are over,
The sun will rise over the horizon.
With new light it bridges
Between what was and what is to be.
With a firm step I cross
And, praying, I ask you to go with me.
I'm not afraid, no matter what.
I have Jesus, my Good Shepherd, my Great Friend!

Chapter 2
The lying messenger

Dear reader,

Throughout this chapter, we will see how a half–truth became a complete lie and was punished with death.

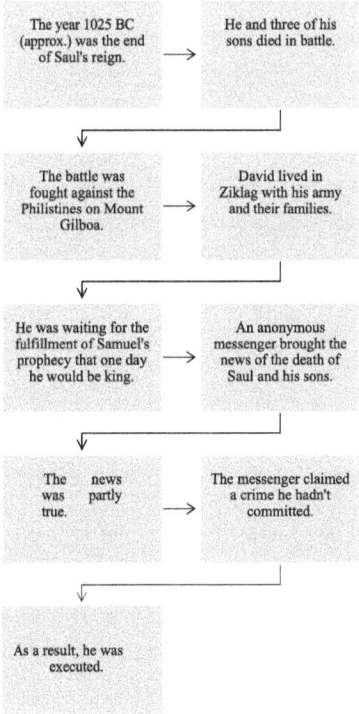

The year 1025 BC (approx.) was the end of Saul's reign. → He and three of his sons died in battle.

The battle was fought against the Philistines on Mount Gilboa. → David lived in Ziklag with his army and their families.

He was waiting for the fulfillment of Samuel's prophecy that one day he would be king. → An anonymous messenger brought the news of the death of Saul and his sons.

The news was partly true. → The messenger claimed a crime he hadn't committed.

As a result, he was executed.

Saul reigned in Israel. The Lord had already decided to take the kingdom away from him and give it to David, "a man after his own heart."

Although Saul did not know that David had already been anointed by the prophet Samuel to be the nation's next king, he hated him deeply and pursued him for nine years with the clear intention of killing him. The close friendship between David and Jonathan, Saul's son, added to the king's jealousy. He understood that it was David, not Jonathan, who would succeed him on the throne. His analysis of the facts and circumstances surrounding David produced in him an insane desire to kill the young warrior, the victor over Goliath, the hero of Israel, about whom the women sang in the streets!

"The women sang as they played, and said,
"Saul has slain his thousands,
And David his ten thousands."

Then Saul became very angry, for this saying displeased him; and he said, "They have ascribed to David ten thousands, but to me they have ascribed thousands. Now what more can he have but the kingdom?" Saul looked at David with suspicion from that day on.

Now it came about on the next day that an evil spirit from God came mightily upon Saul, and he raved in the midst of the house, while David was playing the harp with his hand, as usual; and a spear was in Saul's hand. Saul hurled the spear for he thought, "I will pin David to the wall." But David escaped from his presence twice.

Now Saul was afraid of David, for the LORD was with him but had departed from Saul." (1 Samuel 18:7–12)

Before we go any further, reader, I'd like to briefly catch your attention on a few topics touched on in the text quoted above:

Saul's indignation at the women's singing;

"Rejoice with those who rejoice, and weep with those who weep." (Romans

12:15) – This biblical principle and Christian commandment was not woven into Saul's character. He couldn't stand that David was better than he was. The singing of the anonymous women in the streets of the city was inspired by young David's victory over Goliath, the strongest warrior of the Philistines, who had challenged the ranks of the Israelite army for 40 days, demanding one man – just one – to fight with him. Not one man rose to face the giant! Saul listened to the enemy's insults and challenges for 40 days, but he didn't have the courage, the faith in God, or the confidence to put on his full armor and go into battle. Then along came David, still a teenager, who went to the camp site to bring food to his three older brothers and saw the scene in the theater of war!

He immediately set out to face Goliath. His brothers tried to stop him – to no avail! King Saul tried to stop him – to no avail! Who was this boy? Wasn't he the harpist who sometimes played in the palace to soothe the king?

David's older brother made a point of reminding him that he was nothing more than a shepherd: *"Now Eliab his oldest brother heard when he spoke to the men; and Eliab's anger burned against David and he said, "Why have you come down? And with whom have you left those few sheep in the wilderness? I know your insolence and the wickedness of your heart; for you have come down in order to see the battle."* (1 Samuel 17:28) – This older brother had seen the prophet Samuel anoint David as king one day. He didn't remember it at the time. David was just a boy, he needed to be "set in order", "put in his place", what happened there was a matter for men, not for little boys tending sheep and playing the harp! Or he was trying to protect him from certain death in case he really had to face the giant Goliath.

The prejudice Saul developed against David;

He began to see him as an enemy, an adversary, someone who would draw the attention of the people to himself. Many years earlier, when Saul was anointed king, God's Word tells us that he stood out from the other men from the shoulder up; he was the tallest (I Samuel 10:23). He was also a humble man who barely missed his coronation. He had gone into hiding, and they had to look for him to present him to the people! As the years went by, however, Saul lost some of the qualities that had made him fit to rule. He became impatient and rash. These traits made him a king feared by the people but rejected by the Lord.

When young David appeared on his horizon and defeated Goliath, Saul's envy became obvious, all too obvious. And envy is a deadly emotion! So much so that the Lord Jesus Christ was put to death for envy. In the Gospel of Mark, chapter 15 and verse 10, we read that Pilate *"was aware that the chief priests had handed Him over because of envy."*

Cain killed Abel out of envy. Ten of Joseph's brothers sold him into slavery out of envy! Aaron and Miriam rebelled against their brother Moses out of jealousy and suffered the consequences. In Proverbs 14:30 we read, "Envy is the rottenness of the bones." It undermines thoughts, creates bad feelings, and leads to actions that would never have been done if it weren't for envy! In Saul's case, envy motivated him to seek the death of David, who was his most courageous, trustworthy, fearless, and prudent warrior.

Reader, we must not let envy – of anyone or anything – turn us into Satan's tools to distract us from God's purposes for our lives, let alone to attack the Lord's anointed!

The demonic possession he suffered

Saul opened gaps in his character and behavior through which unclean spirits could torment him! In addition to envy, he was a man without self–control, dominated by a kind of latent anger, fury, and rage. In one such outburst, he tried to kill his own son, Jonathan, simply because he was David's friend!

There is a warning in Ecclesiastes 10:8 that almost passes us by: *"He who digs a pit may fall into it, and a serpent may bite him who breaks through a wall."* Satan is an expert at digging holes and covering them up so that God's people fall into them, get hurt, and never come back. If he finds a small hole in our moral or spiritual life, personality or character, he will try to "bite" us! It will do so mercilessly if he can, if we give him time and space to act!

Regarding anger – which is not always holy, righteous, or godly – Ephesians 4:26–27 gives us the following advice:

"Be angry, and yet do not sin; do not let the sun go down on your anger, and do not give the devil an opportunity."

Satan's influence in the world, the way he tries to steal peace and harmony from God's children, families, and society in general, is not science fiction! It is very real! But he will fall – before the power of the name of the Lord Jesus Christ! All that's needed is for God's children to be vigilant, to stay awake, to pray, and to face him without fear!

Saul didn't protect himself, he didn't guard the walls of his soul – he who had started out so well, so humble and submissive to God, gave way to anger, envy, hatred, resentment and ended his reign very badly. How different it would have been if Saul had recognized David's value as a man and as a warrior in the Lord's wars, his anointing from God to rule, his commitment to the noble cause of defending the nation and strengthening the kingdom, his loyalty to the king himself, against whom he never lifted a finger!

Reader, when you are tempted to get angry, when you feel that anger is too close, that the words you want to say might hurt someone, cry out for help from above! There are words that hurt people close to us to death... Don't let your words hurt anyone.

Let your words be words of blessing, gratitude, recognition, and while you're at it, don't be afraid to praise someone's good work! Be a useful and powerful instrument in God's hands – only in God's hands! Don't harbor feelings of envy or invite pride into your heart. They're Both bad company!

Saul's realization that the Lord was no longer with him, but with David.

We know very well when the Holy Spirit in us is sad when the fire of the Word no longer crackles in our soul, when dryness sets in and the spirit withers. I knew a violin teacher who always knew if his students had studied the lesson during the week. All he had to do was get them to take the violin out of its case. The way they opened the case and picked up the instrument gave them away. For the rest of the lesson, he talked to the students, didn't ask them to play or give them a new piece or a new study. He talked and listened.

I believe that the Heavenly Father does the same with us many times without our understanding what He is saying to us. It's not that the Holy Spirit is abandoning any child of God! It's just that He's hurt, saddened, backed into a corner because He spoke and we didn't obey, He showed us the right way, but we didn't follow it, He reminded us of scriptures, but we didn't want to apply them to our decisions, and so on. He spoke to us with His *"still small voice,"* His *"small, gentle voice,"* but we didn't understand Him. He will continue to wait for us with His patience and His compassionate love, but He will not give us a new mission, a new revelation, or a new lesson as long as we remain far from Him, locked in our arrogance or our stubbornness. What a loss! What a wasted anointing!

Let's go back to the messenger who decided to "exaggerate" his story about the death of King Saul and his three sons:

The Biblical Narrative

Now it came about after the death of Saul, when David had returned from the slaughter of the Amalekites, that David remained two days in Ziklag. On the third day, behold, a man came out of the camp from Saul, with his clothes torn and dust on his head. And it came about when he came to David that he fell to the ground and prostrated himself. Then David said to him, "From where do you come?" And he said to him, "I have escaped from the camp of Israel." David said to him, "How did things go? Please tell me." And he said, "The people have fled from the battle, and also many of the people have fallen and are dead; and Saul and Jonathan his son are dead also." So David said to the young man who told him, "How do you know that Saul and his son Jonathan are dead?" The young man who told him said, "By chance I happened to be on Mount Gilboa, and behold, Saul was leaning on his spear. And behold, the chariots and the horsemen pursued him closely. When he looked behind him, he saw me and called to me. And I said, 'Here I am.' He said to me, 'Who are you?' And I answered him, 'I am an Amalekite.' Then he said to me, 'Please stand beside me and kill me, for agony has seized me because my life still lingers in me.' So I stood beside him and killed him, because I knew that he could not live after he had fallen. And I took the crown which was on his head and the bracelet which was on his arm, and I have brought them here to my lord."

Then David took hold of his clothes and tore them, and so also did all the men who were with him. They mourned and wept and fasted until evening for Saul and his son Jonathan and for the people of the LORD and the house of Israel, because they had fallen by the sword. David said to the young man who told him, "Where are you from?" And he answered, "I am the son of an alien, an Amalekite." Then David said to him, "How is it you were not afraid to stretch out your hand to destroy the LORD's anointed?" And David called one of the young men and said, "Go, cut him down." So he struck him and he died. David said to him, "Your blood is on your head, for your mouth has testified against you, saying, 'I have killed the LORD's anointed.'" (2 Samuel 1:1–16)

A man who didn't know David

A foreigner, a descendant of Amalek,†† came to Ziklag and insisted

---

†† Amalek was the grandson of Esau, the twin brother of Jacob, the son of Isaac and Rebekah. Jacob begat the people of Israel. Esau begat the Edomites. His grandson Amalek fathered the Amalekites, a semi-nomadic people who lived in the desert south of Canaan and were always enemies of the Israelites. The Edomites eventually separated from the Amalekites because they were a cruel people.

on talking to David. He had news to tell him. He wanted to tell him the news firsthand and see how he would react! Perhaps he would even be rewarded for bringing the news! Many messengers have been rewarded for bringing good news!

Curiously, David had just returned from fighting the Amalekites when this man came to him. He brought him news of the death of Saul and his three sons, including Jonathan, David's closest friend.

The Amalekite didn't know David! He didn't know that he had had the opportunity to kill Saul, his adversary, several times, but he had never raised his hand against *"the Lord's anointed,"* as he called him, nor had he allowed any of his men to touch him.

The messenger knew that Saul had been a great enemy of David, that he had persecuted him for years and years, so he decided to "embellish" the story with a few details of his own making:

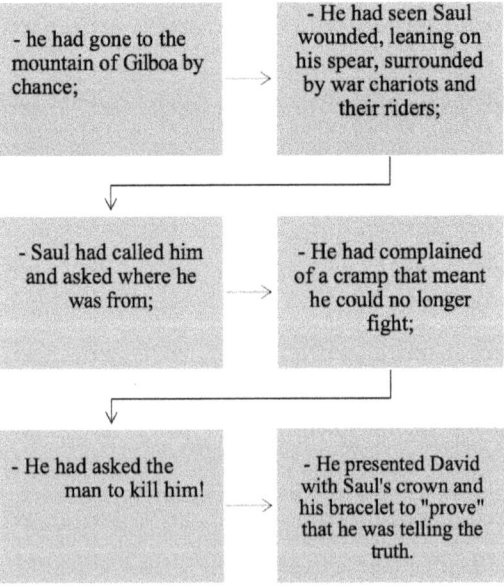

- he had gone to the mountain of Gilboa by chance;

- He had seen Saul wounded, leaning on his spear, surrounded by war chariots and their riders;

- Saul had called him and asked where he was from;

- He had complained of a cramp that meant he could no longer fight;

- He had asked the man to kill him!

- He presented David with Saul's crown and his bracelet to "prove" that he was telling the truth.

To the Amalekite's amazement, David did not rejoice at Saul's death. On the contrary, he tore his clothes and wept, as did the men with him! The messenger must have been confused! Wasn't this good news for David? Why shouldn't it be? Well, there was nothing he could do but wait for everyone to calm down and pull themselves together.

Finally, David called *"the young man who told him the news"*. Not to find out more details about Saul's death, or to give him a reward, but to ask him a blunt question:

"How is it you were not afraid to stretch out your hand to destroy the LORD's anointed?" Without waiting for an answer, David ordered one of his soldiers to execute the messenger on the spot!

"David said to him, "Your blood is on your head, for your mouth has testified against you, saying, 'I have killed the LORD's anointed.'" (2 Samuel 1:16)

This man died because of a lie! If he had only said that Saul and his sons had died in battle against the Philistines, no harm would have come to him. It was just news. Bad news for David, but just news, an accurate account of a tragic event. The Amalekite would not be rewarded, but he would certainly be entitled to some bread, some water, and even some rest before continuing his journey.

However, in his eagerness to please the future king and gain some dividends, he became entangled in a lie that cost him his life.

The power of words

Dear reader, we are not always aware of how much we can build or destroy with words alone! In the case of the Amalekite man, a liar, it was the lie of his words that condemned him. As David said:

"(...) your mouth has testified against you (...)"

Our words have the power to open gaps and to close gaps, to kill and to give life, according to the very explicit text in Proverbs 18:21:

"Death and life are in the power of the tongue, and those who love it will eat its fruit." I invite you to memorize this small portion of Scripture and think about the fruit you can reap from the words of edification, faith, encouragement, and blessing you sow around

you, particularly in the fertile soil of your family!

It's the law of sowing and reaping!

The Lord Jesus Christ left us solemn teachings about the words we speak. Let's look at just these:

- *"For by your words you will be justified, and by your words you will be condemned."* (Matthew 12:37)
- *"But let your statement be, 'Yes, yes' or 'No, no'; anything beyond these is of evil."* (Matthew 5:37)

We have the map. Let's Just follow it!

The Lord Jesus Christ was also very clear about Satan and lies. He once said these words to a group of men who were confronting Him:

"You are of your father the devil, and you want to do the desires of your father. He was a murderer from the beginning, and does not stand in the truth because there is no truth in him. Whenever he speaks a lie, he speaks from his own nature, for he is a liar and the father of lies." (John 8:44)

In this short passage, we can highlight several aspects of Satan's personality that are revealed to us:

- He has followers who want to please him;
- He has his own desires, plans and schemes;
- He is a murderer;
- He is intrinsically a liar;
- He spawned the sin of lying, and he spawns lies every day: he is skilled in disguise and deception!

On the other hand, in the same chapter 8 of the Gospel of John, the Lord Jesus Christ tells us:

"If you continue in My word, then you are truly disciples of Mine; and you will know the truth, and the truth will make you free." (John 8:31–32)

He is the way, the truth and the life! He is the living Word of God! The apostle John wrote about Him:

"In the beginning was the Word, and the Word was with God, and the Word was God. He was with God in the beginning. Through him all things were made; without him nothing was made that has been made. In him was life, and that life was the light of all mankind." (John 1:1–4, NIV)

The Word called into being everything that did not yet exist!

"By faith we understand that the worlds were prepared by the word of God, so that what is seen was not made out of things which are visible." *(Hebrews 11:3)*

When we read the account of the creation of the world in Genesis 1, we marvel at the fact that God spoke and everything came into being! Too often, however, we forget that this is our God, that His power is the same, that His Word continues to create new life, to save, to heal, to bless, to guide, to teach, to build…

In His infinite goodness, the Lord created us with the ability to use words, not only to understand each other, but also to speak blessings, words of faith, edification, transformation, hope, kindness, love and grace over our own lives and the lives of others!

But let's remember that lies are a powerful weapon in the hands of Satan in the world of darkness. He has used it against mankind since the Garden of Eden and continues to use it every day! He uses it to control nations and governments, businesses and industries, the media, marketing and advertising, philosophies of life, religions, human minds individually and collectively, and influences countless other areas that we don't even know about!

There is only one way to erect barriers and break this power on a personal level: disarm it! Use God's truth against every form of lie, no matter how disguised, no matter how much "others" act the same way, because "everyone will give account of himself to God." (Romans 14:12). To succeed, we must know the Word of Truth and the Lord of the Word and the Truth!

A moment alone with God

REFLECTION ON:

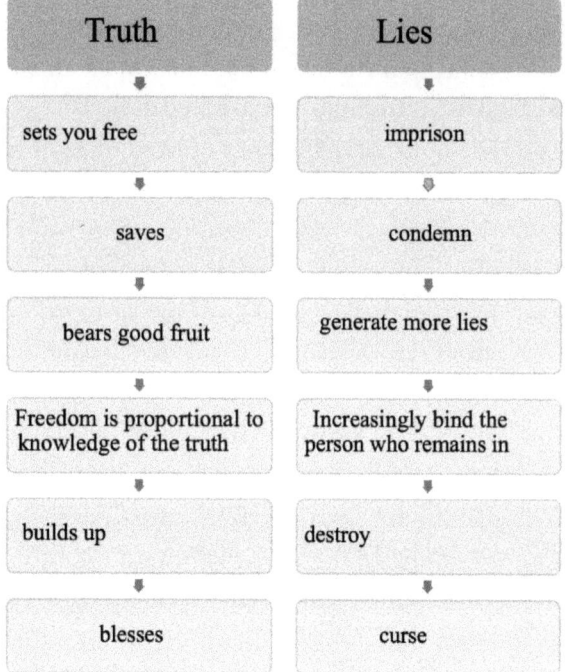

Truth	Lies
sets you free	imprison
saves	condemn
bears good fruit	generate more lies
Freedom is proportional to knowledge of the truth	Increasingly bind the person who remains in
builds up	destroy
blesses	curse

PRAYER

Dear Lord,

Today I choose words of blessing and reject all words of cursing. I choose truth and reject lies.

I understand that through words I can open hostilities or promote peace. I choose to sow words of peace to reap more peace with joy. I choose to distance myself from people who encourage me to use bad words, to be slanderous and ungrateful, for it is written that *"Bad company corrupts good morals."* (1 Corinthians 15:33)

I choose to use the Word of God as a weapon of defense against the wiles of Satan and as a weapon of attack against the ungodly thrust of the enemy!

I choose to use the Word under the control and direction of the Holy Spirit, in the power of the name of the Lord Jesus Christ, to destroy strongholds and to pull down sophisms, as it is written in 2 Corinthians 10:4.

I choose to speak the truth, without taking away or adding anything, without being biased or partial, even when I am tempted to deviate "a little" for my own benefit.

"Create in me a clean heart, O God, and renew a steadfast spirit within me." (Psalm 51:10).

In the name of Jesus I pray.